6/15/05 amazon 2

# RACHEL CARSON

## Recent Titles in Greenwood Biographies

Colin Powell: A Biography
*Richard Steins*

Pope John Paul II: A Biography
*Meg Greene Malvasi*

Al Capone: A Biography
*Luciano Iorizzo*

George S. Patton: A Biography
*David A. Smith*

Gloria Steinem: A Biography
*Patricia Cronin Marcello*

Billy Graham: A Biography
*Roger Bruns*

Emily Dickinson: A Biography
*Connie Ann Kirk*

Langston Hughes: A Biography
*Laurie F. Leach*

Fidel Castro: A Biography
*Thomas M. Leonard*

Oprah Winfrey: A Biography
*Helen S. Garson*

Mark Twain: A Biography
*Connie Ann Kirk*

Jack Kerouac: A Biography
*Michael J. Dittman*

Mother Teresa: A Biography
*Meg Greene*

Jane Addams: A Biography
*Robin K. Berson*

# RACHEL CARSON

## A Biography

Arlene R. Quaratiello

GREENWOOD BIOGRAPHIES

**GREENWOOD PRESS**
WESTPORT, CONNECTICUT · LONDON

**Library of Congress Cataloging-in-Publication Data**

Quaratiello, Arlene Rodda.
    Rachel Carson : a biography / Arlene R. Quaratiello.
        p. cm.—(Greenwood biographies, ISSN 1540–4900)
    Includes bibliographical references (p.   ).
    ISBN 0–313–32388–7
    1. Carson, Rachel, 1907–1964. 2. Biologists—United States—Biography.
3. Environmentalists—United States—Biography. I. Title. II. Series.
QH31.C33Q37 2004
574'.092—dc22          2004010980

British Library Cataloguing in Publication Data is available.

Library of Congress Catalog Card Number: 2004010980
ISBN: 0–313–32388–7
ISSN: 1540–4900

First published in 2004

Greenwood Press, 88 Post Road West, Westport, CT 06881
An imprint of Greenwood Publishing Group, Inc.
www.greenwood.com

Printed in the United States of America

The paper used in this book complies with the
Permanent Paper Standard issued by the National
Information Standards Organization (Z39.48–1984).

10  9  8  7  6  5  4  3  2  1

# CONTENTS

*Photo essay follows page 62.*

# SERIES FOREWORD

In response to high school and public library needs, Greenwood developed this distinguished series of full-length biographies specifically for student use. Prepared by field experts and professionals, these engaging biographies are tailored for high school students who need challenging yet accessible biographies. Ideal for secondary school assignments, the length, format and subject areas are designed to meet educators' requirements and students' interests.

Greenwood offers an extensive selection of biographies spanning all curriculum related subject areas including social studies, the sciences, literature and the arts, history and politics, as well as popular culture, covering public figures and famous personalities from all time periods and backgrounds, both historic and contemporary, who have made an impact on American and/or world culture. Greenwood biographies were chosen based on comprehensive feedback from librarians and educators. Consideration was given to both curriculum relevance and inherent interest. The result is an intriguing mix of the well known and the unexpected, the saints and sinners from long-ago history and contemporary pop culture. Readers will find a wide array of subject choices from fascinating crime figures like Al Capone to inspiring pioneers like Margaret Mead, from the greatest minds of our time like Stephen Hawking to the most amazing success stories of our day like J. K. Rowling.

While the emphasis is on fact, not glorification, the books are meant to be fun to read. Each volume provides in-depth information about the subject's life from birth through childhood, the teen years, and adulthood. A thorough account relates family background and education, traces personal and professional influences, and explores struggles, accomplishments, and contributions. A timeline highlights the most significant life events against a historical perspective. Bibliographies supplement the reference value of each volume.

# INTRODUCTION

Rachel Carson's obituary in the *New York Times* described her as "a small, solemn-looking woman with the steady forthright gaze of a type that is sometimes common to thoughtful children who prefer to listen rather than to talk" (*New York Times* 1964: 25). But when the reserved and unpretentious author of *Silent Spring* as well as three best-selling books about the sea did express herself, society paid close attention. Carson introduced concepts relating to the environment and conservation that were virtually ignored by most people during the mid–twentieth century and brought terms such as "interdependence" and "the balance of nature" into common usage. She is considered by many to be "the fountainhead of the modern environmental movement" (Lear 1999: 474).

Rachel Carson is remembered almost exclusively for her last book, *Silent Spring*. Years before this magnum opus, however, the publication of Carson's sea trilogy—*Under the Sea-Wind, The Sea Around Us,* and *The Edge of the Sea*—made her a celebrity. These books "stirred people to love the sea because of its beauty for which she was their eyes, for its mystery of which she was the oracle, and for its cadence and sound for which she was its voice" (Hynes 1989: 35). Although she was a renowned marine biologist and the best-selling author of books about the aquatic environment and its creatures, she never saw the ocean during her childhood. Not until the summer after her graduation from college did she first behold the sea.

Rachel always had a great appreciation for the natural world as well as a deep respect for the interconnectedness of all creatures. Her attitude exemplified what the renowned humanitarian Dr. Albert Schweitzer called "reverence for life," a philosophy that espouses respect for the interdependence of all living things and considers "all life sacred, including forms of life that from the human point of view may seem to be lower than ours" (Schweitzer 1965: 47). When Carson received the Schweitzer Medal of the Animal Welfare Institute in 1963, she paraphrased the words of the award's namesake in her acceptance speech, saying "Dr. Schweitzer has told us that we are not being truly civilized if we concern ourselves only with the relation of man to man. What is important is the relation of man to all life" (quoted in Brooks 1972: 316).

Carson believed that "a large share of what's wrong with the world is man's towering arrogance—in a universe that surely ought to impose humility, and reverence" (quoted in Freeman 1995: 241). Her body of work, including not only her best-selling books but also her magazine articles, government brochures, speeches, and virtually everything else she wrote, reflects her belief that all life on earth is interrelated and that human beings should consider themselves part of the natural environment, not masters of it. Her writing "became a catalyst for change. A debate had begun: a reverence for life versus a reverence for power" (Williams 1992: 105–06).

Rachel's own "reverence for life" is evident in both her writing and her actions. In a field notebook that she kept while visiting Saint Simons Island off the coast of Georgia in the spring of 1952, she recounted an episode that revealed her respect for all living things. While walking along the beach, she noticed a small dog far out on the flats jumping around in the tide pools, wagging his tail in contentment, oblivious to the incoming tide. As the waves continued to come closer, Rachel became very concerned that the dog would be trapped and drowned, so she went out to pick him up and brought him to safety.

In *The Edge of the Sea*, Rachel told a story about a starfish that she had taken to study early one morning at low tide. That night, when the tide was out again, she brought the starfish back to the place she had taken it from, knowing that it had to be returned to its habitat if it was to survive. Each creature that Rachel collected was not only a specimen to her, but a link in the vast chain of life and, as such, something to be respected. When she first saw a particularly beautiful West Indian basket fish, she "stood beside it, lost to all but its extraordinary and somehow fragile beauty. I had no wish to 'collect' it; to disturb such a being would have seemed a desecration" (Carson 1955: 225).

Perhaps the strongest indication of Carson's deep respect for all living creatures was her struggle to write *Silent Spring* despite failing health and intense criticism. Through her research, she came to believe that all life was threatened by the misuse of pesticides and that the natural beauty she cherished was being destroyed. No matter what the cost to her own health and reputation, she knew that she had to complete *Silent Spring*.

Although Rachel was not an overtly religious person, neither was she an atheist, and she did acknowledge God as the maker of the natural wonders of the universe. When her mother told her that God was the creator of the world, Rachel said, "Yes, and General Motors created my Oldsmobile. But *how* is the question" (quoted in Anticaglia 1975: 212). There were many readers of *The Sea Around Us* who took issue with Carson's view of evolution as expressed in her book because it seemed to deny the existence of God. In response to a letter from attorney James Bennet expressing his objections to her book, she wrote:

> It is true that I accept the theory of evolution as the most logical one that has ever been put forward to explain the development of living creatures on this earth. As far as I am concerned, however, there is absolutely no conflict between a belief in evolution and a belief in God as the creator. Believing as I do in evolution, I merely believe that is the method by which God created and is still creating life on earth. And it is a method so marvelously conceived that to study it in detail is to increase—and certainly never to diminish—one's reverence and awe both for the Creator and the process. (quoted in Lear 1997: 227)

In a letter to her friend Dorothy Freeman, Rachel alluded to her beliefs, stating, "I am sure, there is a great and mysterious force that we don't, and perhaps never can understand" (quoted in Freeman 1995: 241). In another letter, she elaborated on this thought, explaining, "Because I cannot understand something doesn't mean it doesn't exist" (447).

According to editor Paul Brooks, Rachel had "a youthful enthusiasm, a sense of adventure that, to the end of her life, turned the humblest trip into a voyage of discovery" (Brooks 1972: 12). She herself called this characteristic a "sense of wonder" and elaborated on it in an article she wrote for *Woman's Home Companion* titled "Help Your Child to Wonder." This attitude was basically an intense appreciation of nature that reflected her deep respect for all living things. Rachel felt that a "sense of wonder" was important because those who have it "are never alone or weary of life." They have "inner contentment" and "find reserves of strength that will endure as long as life lasts" (Carson 1956: 48). She once observed, "There is one quality that characterizes all of us who deal with the sciences of the earth and its life—we are never bored. We can't be. There is always something new to be investigated. Every mystery solved brings us to the threshold of a greater one" (quoted in Lear 1998: 159). Rachel's particularly strong "sense of wonder" was reflected in all of her writing as she shared the fascination she experienced when studying living things.

Carson's own "reverence for life" and her "sense of wonder" helped her to surmount numerous obstacles in the pursuit of her life's purpose. She endured financial hardships that prevented her from being a full-time writer and also had substantial family responsibilities including the care of her mother, her nieces, and her great-nephew. She wrote to Dorothy Freeman, "my great problem was how to be a writer and at the same time a member of my family.... It is that conflict that just tears me to pieces" (quoted in Freeman 1995: 98). She also suffered the consequences of being a woman in what was then a "man's world" and was criticized as being a hysterical spinster after the publication of *Silent Spring*.

Because Carson wrote for general readers rather than scientists, she was able to share her philosophy and attitudes with a much larger audience. She was criticized by some colleagues for popularizing science, but she felt that all people could benefit from understanding scientific principles, affirming, "Knowledge of the facts of science is not the prerogative of a small number of men, isolated in their laboratories, but belongs to all men, for the realities of science are the realities of life itself" (quoted in Lear 1998: 165). She believed that an appreciation of nature was something that anyone could possess, writing, "The pleasures, the values of contact with the natural world, are not reserved for the scientists. They are available to anyone who will place himself under the influence of a lonely mountain top—or the sea—or the stillness of a forest; or who will stop to think about so small a thing as the mystery of a growing seed" (160).

# TIMELINE OF EVENTS IN THE LIFE OF RACHEL CARSON

| | |
|---|---|
| 27 May 1907 | Birth of Rachel Carson. |
| September 1913 | Entered School Street School. |
| September 1918 | "A Battle in the Clouds" published in *St. Nicholas Magazine*. |
| September 1923 | Entered Parnassus High School. |
| May 1925 | Graduated first in class from Parnassus High School. |
| September 1925 | Entered Pennsylvania College for Women as an English major. |
| January 1928 | Officially changed her major to biology. |
| 10 June 1929 | Graduated magna cum laude from Pennsylvania College for Women with a B.A. in biology. |
| August 1929 | Arrived at the Marine Biological Laboratory in Woods Hole, Massachusetts, for a six-week course as a "beginning investigator." |
| October 1929 | Began graduate studies at Johns Hopkins University. |
| Spring 1930 | The Carson family moved to Baltimore, Maryland, to live with Rachel. |
| Summer 1930 | Began teaching summer school at Johns Hopkins, where she taught until 1936. |
| September 1931 | Began teaching at the University of Maryland. |
| 14 June 1932 | Awarded a master's degree in marine zoology by Johns Hopkins. |
| Winter 1934 | Formally withdrew from doctoral studies. |

| | |
|---|---|
| 6 July 1935 | Death of Rachel's father, Robert Carson. |
| October 1935 | Began part-time, temporary work at the Bureau of Fisheries writing "Romance Under the Waters" scripts. |
| January, May 1936 | Took civil service exams. |
| 1 March 1936 | First newspaper article appeared in the *Baltimore Sun*, "It'll Be Shad-Time Soon." |
| 17 August 1936 | Began working full time at the Bureau of Fisheries as a junior aquatic biologist in the Division of Scientific Inquiry. |
| January 1937 | Death of Rachel's sister Marian. |
| September 1937 | "Undersea" published in the *Atlantic Monthly*. |
| 1 November 1941 | Publication of *Under the Sea-Wind*. |
| May 1942 | Promoted to assistant aquatic biologist in the Division of Information at the Fish and Wildlife Service (FWS). |
| August 1942 | Relocated temporarily to Chicago; began work on *Food from the Sea* brochures. |
| May 1943 | Promoted to associate aquatic biologist in the FWS Office of the Coordinator of Fisheries; moved back to Maryland. |
| 1945 | Promoted to aquatic biologist in the FWS Division of Information; began work on *Conservation in Action* booklets. |
| Summer 1946 | Spent a month in Boothbay Harbor, Maine, to celebrate her ten years with the Fish and Wildlife Service. |
| 1946 | Promoted to information specialist in the Division of Information. |
| July 1949 | First and only underwater diving experience in Florida; ten-day voyage aboard the *Albatross III* to Georges Bank. |
| 1949 | Promoted to biologist and chief editor in the Division of Information; awarded a Saxton Fellowship. |
| September 1950 | Publication of "The Birth of an Island" in the *Yale Review*, which won the Westinghouse Science Writing Prize. |
| March 1951 | Awarded a Guggenheim Fellowship. |
| June 1951 | Three-part condensation of *The Sea Around Us* published in the *New Yorker*. |
| 2 July 1951 | Publication of *The Sea Around Us*. |

| | |
|---|---|
| 9 September 1951 | *The Sea Around Us* hit number one on the *New York Times* best-seller list. |
| 1951 | The National Book Award given to *The Sea Around Us*. |
| 9 January 1952 | Became the first woman to receive the Henry G. Bryant Medal. |
| 18 February 1952 | Birth of Roger Christie, Carson's grandnephew. |
| 13 April 1952 | *Under the Sea-Wind* was rereleased and became a bestseller. |
| May 1952 | Received honorary doctorate from Pennsylvania College for Women. |
| June 1952 | Resignation from the Fish and Wildlife Service; received honorary doctorates from Drexel Institute of Technology and Oberlin College. |
| 1952 | Awarded the John Burroughs Medal and the Garden Club of America's Frances K. Hutchinson Medal. |
| January 1953 | Received the New York Zoological Society's Gold Medal. |
| June 1953 | Received honorary doctorate from Smith College. |
| Summer 1953 | Spent the summer in Southport Island, Maine, in her newly built cottage. |
| May 1954 | Awarded the Silver Jubilee Medal from the Limited Editions Club. |
| August 1955 | Appearance of the *New Yorker*'s condensed version of *The Edge of the Sea*. |
| 26 October 1955 | Publication of *The Edge of the Sea*. |
| November 1955 | Elected an honorary fellow of the Museum of Science in Boston. |
| 11 March 1956 | TV documentary "Something about the Sky," written by Carson, aired on CBS. |
| June 1956 | Received Achievement Award from the American Association of University Women. |
| July 1956 | "Help Your Child to Wonder" appeared in *Woman's Home Companion*. |
| 30 January 1957 | Death of Rachel's niece Marjorie. |
| July 1958 | "Our Ever-Changing Shore" appeared in a special edition of *Holiday*. |
| December 1958 | Death of Rachel's mother, Maria Carson. |
| June 1962 | Appearance of the *New Yorker*'s condensed version of *Silent Spring*. |

| | |
|---|---|
| 27 September 1962 | Publication of *Silent Spring*. |
| January 1963 | Awarded the Albert Schweitzer Medal of the Animal Welfare Institute. |
| 3 April 1963 | CBS aired "The Silent Spring of Rachel Carson." |
| 15 May 1963 | Release of the President's Science Advisory Committee report "Use of Pesticides," which vindicated Carson. |
| 4 June 1963 | Gave testimony before the Senate subcommittee headed by Senator Abraham Ribicoff, which investigated the misuse of pesticides. |
| December 1963 | Awarded the Audubon Medal; presented with the American Geographical Society's Cullum Medal; inducted as a member in the American Academy of Arts and Letters. |
| 1963 | Received the Conservationist of the Year Award from the National Wildlife Federation and the National Council of Women's first Woman of Conscience award. |
| 14 April 1964 | Death of Rachel Carson. |
| 1965 | *A Sense of Wonder* published; the Rachel Carson Council founded. |
| 22 April 1970 | First Earth Day. |
| 1970 | Formation of the Environmental Protection Agency; dedication of the Rachel Carson National Wildlife Refuge in Maine. |
| 1972 | Environmental Protection Agency banning of the pesticide DDT in the United States. |
| 1980 | The Presidential Medal of Freeman awarded posthumously by Jimmy Carter. |
| 1981 | Rachel Carson postage stamp issued by the U.S. Postal Service. |

# Chapter 1

# LEARNING AND LOVING NATURE

From her earliest recollections, Rachel Carson possessed a special affinity for nature. She spent a great deal of her childhood exploring the woods on her family's property and enjoyed identifying birds, insects, and flowers. Born on May 27, 1907, in Springdale, Pennsylvania, Rachel came into the world at a time when many rural areas such as her hometown were succumbing to industrialization. Although few people at the time were concerned about environmental issues, this development posed a great threat to the natural world that Rachel loved.

Remarkably, it was the ocean that most strongly attracted Rachel, even though she lived hundreds of miles from the Atlantic coast and had never even visited the ocean as a child. Some neighbors speculated that a fossilized shell that Rachel found near her home provoked this infatuation with the sea. A curious child, she read every book she could find about the ocean environment that fascinated her so that she could more clearly picture it in her dreams.

## RACHEL'S PARENTS

Rachel's father, Robert Warden Carson, had grown up in Allegheny City, across the river from Pittsburgh, Pennsylvania. Born in 1864, he was the eldest of the six children of James and Ellen Carson. His parents had emigrated from Ireland, and his father worked as a carpenter. In 1894, when Robert was thirty years old, he performed with his Presbyterian church quartet at a choral social in Canonsburg, Pennsylvania. There he met Maria McLean, a soloist for the Washington Quinette Club from Washington, Pennsylvania.

Maria Frazier McLean was born in 1869 in Cleveland, Ohio, and moved several times as a child to the different towns where her father, Daniel M. B. McLean, served as a Presbyterian minister. After living in Allegheny City, the McLean family moved back to Cleveland, and then to Canonsburg, Pennsylvania. When Maria was eleven, her father, only forty years old, died of tuberculosis. After his death, she moved with her mother, Rachel, and sister, Ida, to Washington, Pennsylvania, Rachel McLean's hometown. Maria was a talented pianist and singer who studied classics at the elite Washington Female Seminary, a Presbyterian school. She graduated with honors in Latin in 1887 and then became a teacher.

Despite Robert Carson's lack of schooling and lower social class, he married Maria in June 1894. Maria had to give up her teaching career after marrying Robert at the age of twenty-five because married women during that era were not permitted to teach. The Carsons lived in Canonsburg where daughter Marian was born in 1897, and Robert, Jr., two years later. They moved to Springdale, twenty miles north of Pittsburgh in the lower Allegheny valley, in 1900 after Robert Carson had bought sixty-five wooded acres on the western outskirts of the town on the shores of the Allegheny River. Seven years later, Rachel was born.

Robert Carson held a variety of jobs but was never successfully employed. At various times, he worked as a salesman for the Great American Insurance Company, an electrician at the Harwick Mine, and a part-time employee of the West Penn Power Company. His extended family included several farmers, and he considered carrying on that tradition also, but the land he purchased in Springdale never became a true working farm. With dreams of becoming a developer, he dabbled in real estate and hoped to augment his income by selling parcels of the family farm.

## MAJOR INFLUENCES ON RACHEL

Rachel's mother, having been raised from the age of eleven in an all-female household, was an opinionated and domineering figure who had a stronger influence on Rachel than her father, an emotionally distant person who was frequently away from home on business trips. Rachel and her mother developed a close relationship, and they would live together until Maria's death in 1958. Paul Brooks, Rachel's editor and friend, asserted, "Maria Carson was undoubtedly the strongest single influence in her daughter's life" (Brooks 1972: 242). Rachel was born when her mother was thirty-eight. Because her other children, Marian and Robert, were considerably older than Rachel, Maria could focus her attention on her youngest child. She instilled in Rachel an appreciation of nature as well as a love of literature and music.

Maria Carson enjoyed natural history, botany, and bird-watching and possessed a great respect for nature that she passed along to her daughter. She would never even kill an insect in the house and would only grudgingly cook the rabbits that son Robert brought home after hunting. She was interested in the nature-study movement that was popular around the turn of the century. This movement, which had a spiritual element, was in part a reaction against industrialization, and its advocates dedicated themselves to instilling in children a love of nature so that they would not be alienated from it in an increasingly urban world. From the time Rachel was a year old, Maria would take her outside every day to explore the family property and teach her the outdoor lessons from Anna Botsford Comstock's *Handbook of Nature Study*, published in 1911, a popular book at the time for proponents of the nature-study movement.

Rachel's mother also shared with her daughter a love of books and writing. Some of Rachel's favorite stories featured animals that possessed human qualities. She especially loved the beautifully illustrated tales of Beatrix Potter, which featured such anthropomorphic characters as Peter Rabbit, Benjamin Bunny, and Squirrel Nutkin. Carson biographer Philip Sterling speculated that these stories provided "a mirror of her feelings about the fields and woods, about the small creatures, tame and wild, who were part of her personal world—of her environment." He went on to theorize that this literature may have suggested to her that "there may be some great mysterious unity in all of nature; that everything in the world may somehow be connected with everything else" (Sterling 1970: 22–23). Rachel's childhood respect for the interconnectedness of all living things was an attitude that would guide her throughout her life.

Rachel also enjoyed the works of Ernest Thompson Seton, the founder of the Boy Scouts of America, who wrote numerous children's books about nature. She read the novels of Gene Stratton Porter, an "apostle of the nature-study movement who believed that through nature a child was led to God" (Lear 1997: 17). She was also fascinated by any literature pertaining to the sea, including the works of such authors as Herman Melville, Joseph Conrad, and Robert Louis Stevenson. Although Rachel later described herself as "rather a solitary child" who read a great deal, she was certainly not antisocial (quoted in Lear 1998: 148). She enjoyed spending time with friends Irene Mills, Charlotte Fisher, and Mildred Krumpe, who shared her interest in books.

Besides her mother and the books she read, a major influence on Rachel was the environment in which she grew up. Although the family farm was not a real farm, it did have a pig, cows, chickens, and horses, as well as woods, streams, and fields to explore. There was also an apple orchard that provided a site for frequent picnics and town gatherings. Rachel spent many happy hours outdoors, growing to love and respect the natural world and its creatures. But Springdale slowly succumbed to industrialization and grew in population. In 1900, there were approximately twelve hundred residents, but by 1910 there were twice that number. Coal mining was an important industry, and, at one point, the coal company wanted to extend its mine shaft under the Carson property, but Robert Carson refused because it might have a negative effect on future property value. Iron and steel were also produced in the area, and oil and timber were transported on the Allegheny River, while a glue factory sent noxious fumes into the air. The town later became situated between two power plants, the West Penn Power Company and the Duquesne Light Company.

## SCHOOL DAYS AND EARLY WRITING ENDEAVORS

In 1913, Rachel began attending Springdale's elementary school, School Street School, less than a mile from her home. She was always an excellent student, earning mostly A's despite long absences. After contracting scarlet fever as a young child, she was never as strong as the other children, and her mother often kept her out of school if another student was sick to prevent her daughter from catching an illness. During her time spent at home, including an entire month of fourth grade, her mother, a former teacher, tutored her quite adequately.

Rachel could "remember no time, even in earliest childhood, when I didn't assume I was going to be a writer" (quoted in Lear 1998: 148). She didn't know why, especially since there were no writers in her family. She simply thought it would be fun to tell stories like the people who wrote the books that she read. She began writing poetry at the age of eight, and in the spring of 1918, when she was ten years old, she submitted a story to *St. Nicholas Magazine,* a popular children's periodical that promulgated the values of the nature-study movement and featured the work of young writers in the "St. Nicholas League" section. Such famous authors as William Faulkner, F. Scott Fitzgerald, Edna St. Vincent Millay, and Eudora Welty had previously been published in the "League."

Rachel had read the stories in the magazine for years and finally submitted her own 253-word piece. Her story, "A Battle in the Clouds," was inspired by a tale told to her by her brother, who was in the U.S. Army Air Service during World War I. In it, a Canadian flyer whose plane is shot at by the Germans averts a crash by hanging from a wing and balancing the plane. After impressing the enemy with his heroics, he is allowed to land safely. The story, which appeared in the September 1918 issue, won the League's silver badge.

Rachel wrote three stories that appeared in *St. Nicholas Magazine* during 1919. "A Young Hero" was published in January, "A Message to the Front" appeared the following month and won the gold badge, and "A Famous Sea-Fight" was included in the August issue. All of these stories focused on military themes, reflecting the impact that World War I, which had recently ended, had had upon the young Rachel. She became an "Honor Member" in the League and received $10 in cash. This award thoroughly convinced her that she would become a professional writer. Rachel considered her career to have truly started when the magazine paid her a penny a word for an essay she wrote about St. Nicholas. Her final piece for the magazine was "My Favorite Recreation," published in July 1922, a short descriptive essay that concerned her interest in bird-watching in the hills of Pennsylvania with her dog.

Rachel was much more ambitious than her older siblings. Her sister, Marian, never finished high school, choosing instead to work as a stenographer. Marian was married in 1915 at age eighteen to Lee Frank Frampton and continued to live for a brief time in the small family home. After a tumultuous marriage, she divorced her husband in 1918 and worked as a bookkeeper at the largest employer in the area, West Penn Power Company. In 1920, she married again to a stenographer named Burton P. Williams, with whom she had two daughters, Virginia and Marjorie. Rachel's brother, Robert, who also did not finish high school, joined the Army Air Service in 1917 and fought in France during World War I. When he returned to Springdale after the war, he was eventually hired as an electrician's assistant at the power company.

The studious Rachel, however, continued to excel in high school, maintaining an A average. Springdale did not have its own secondary school, and the Carsons could not afford the train fare to the school in a nearby town, so Rachel continued to be tutored at School Street School in Springdale for the first two years of high school, attaining a 93.7 average. After her sophomore year, she transferred to Parnassus High School in New Kensington two miles away and maintained her high marks. Aside from academics, she participated in basketball and field hockey, but it was difficult to get involved in activities and friendships because of transportation problems. She graduated first in her class from Parnassus High in May 1925.

## HIGHER EDUCATION

In the fall of 1925, Rachel entered Pennsylvania College for Women (PCW). The school, which sat on a hill northeast of the business district of Pittsburgh, was the only one that she and her mother had considered because of its proximity, small size, reputation, and Christian affiliation. Rachel wanted to become a writer and so declared her major as English, believing that this course of study would provide the proper background and training for her chosen career.

Like most college women of the time, many of the other students considered their college experience merely a transition between living with their parents and starting their own families as educated wives and mothers. Some might go on briefly to work in such fields as teaching or nursing, but students such as Rachel who planned on lifelong professional careers were unusual. Rachel was a gifted student and had been awarded a $100 scholarship by the school after winning an academic competition and another scholarship for the same amount awarded by the State Department of Instruction for excelling on the annual state scholarship exam. These funds, however, would cover only a fraction of the $800 annual tuition, plus room, board, and other expenses. She was such a prized student that the administration, including President Cora Helen Coolidge and Dean Mary Helen Marks, didn't want her to drop out because of a lack of money. Some additional assistance was discreetly provided by the school from some wealthy friends of the president. The Carsons hoped to raise the rest of the money by selling off plots of land, but they were only able to raise a small amount. Rachel took out a number of loans and at one point used parcels of her father's land as collateral.

Rachel excelled as an English major and was taught by Professor Grace Croff, a new assistant professor and instructor of freshman composition. Professor Croff was a demanding teacher and became both Rachel's mentor and her friend during the second semester of her freshman year. In her first essay, titled "Who I Am and Why I Came to PCW," Rachel expressed her love of the outdoors and nature and revealed that she considered herself to be an ambitious idealist. In this composition she wrote, "Sometimes I lose sight of my goal, then again it flashes into view, filling me with a new determination to keep the 'vision splendid' before my eyes" (quoted in Lear 1997: 32).

Rachel became a frequent contributor to the *Arrow*, the school's newspaper, and was also a member of Omega, the literary society. She had some of her stories published in the *Englicode*, the literary supplement to the paper, including "The Master of the Ship's Light," which appeared at the end of her freshman year. In this story Rachel depicted the sea in great detail, writing, for example, "Over the surface of the long lazy swells that rolled in on the shallow beach, played dark formless shadows or patches of white foam, betraying the menacing reefs beneath. When the icy winds swept down from the Straits, towering waves beat upon the coast with uncontrollable fury, and the booming of the breakers resounded for miles" (quoted in Sterling 1970: 45). Considering that Rachel had never even seen the ocean, her descriptions are quite remarkable. This story was originally written for a class taught by Croff who, in a remark that foreshadowed the writing style that would make Rachel famous, commented, "Your style is so good because you have made what might be a relatively technical subject very intelligible to the reader" (quoted in Lear 1997: 33–34).

Carson wrote a number of other noteworthy stories for the *Englicode*. "Why I Am a Pessimist" was a story told from the viewpoint of a housecat who desires to be an equal member of the household. "Keeping an Expense Account" was an autobiographical essay about a task she loathed—tracking her finances. In "The Golden Apple," a new version of a Greek myth, Rachel wondered why women depended on fathers and husbands and other men to make decisions for them.

Rachel was also a talented poet as demonstrated by the following triolet, an eight-line poem with a difficult rhyming scheme:

> Butterfly poised on a thistle's down.
> Lend me your wings for a summer's day.
> What care I for a kingly crown?
> Butterfly poised on a thistle's down.
> When I might wear your gossamer gown
> And sit enthroned on an orchid spray.
> Butterfly poised on a thistle's down.
> Lend me your wings for a summer's day! (quoted in Lear 1997: 42)

This poem was written for one of the courses that Grace Croff taught during the first semester of Rachel's junior year at a time when she was struggling with difficult decisions in her life. During her senior year, she submitted a number of her poems to some national magazines. Unfortunately, nothing remains of this poetry, and only the rejection letters she received from the magazine editors attest to its existence.

## SOCIAL LIFE IN COLLEGE

Although she usually preferred studying to socializing, Rachel enjoyed participating on the basketball, baseball, and field hockey teams and also pursued extracurricular activities such as swimming, tennis, hiking, and riding. She did have a few good friends including her best friend Marjorie Stevenson, a day student and history major, and biology majors Mary Frye and Dorothy Thompson, who were a year behind her.

Living in a dormitory at the school, which was only sixteen miles from her home, was Rachel's first experience of being away from her family. Her mother, who continued to be a strong influence in her life, came to visit every other weekend, and Rachel went home on alternate weekends, making it difficult to cultivate strong friendships. Her mother, who was nicknamed "the commuter" by some of the students with whom Rachel lived, had never had the opportunity to attend college herself, so she enjoyed the experience vicariously.

Due to the ambition that isolated her, Rachel never quite fit in socially at PCW, but because her goals were academic, this was of no concern to her. Classmate Mary Kolb commented about Rachel, "Though she was much more of a scholar than the rest of us and in a way withdrawn, she entered into things with great spirit. When you asked her to do something, she did it wholeheartedly—if she wanted to do it" (quoted in Sterling 1970: 43). Philip Sterling suggested that Rachel "seemed to need no one's good opinion except her own. Her manner was friendly though not warm. She was soft-spoken but not meek" (Sterling 1970: 43). Carson biographer Linda Lear saw her in a similar way, writing, "Self-confident about her intellectual abilities and solitary by nature, Rachel had few social graces and little understanding of how to interact in a wider society. She was fiercely determined to become all that she could be for herself. She also had a vision, not yet articulated, an inchoate sense of some special calling that awaited her" (Lear 1997: 26).

Rachel did have a subtle sense of humor and enjoyed a prank now and then. One evening she was in the lab with Dorothy Thompson, and they noticed that the alcohol supply was almost gone. Use of the alcohol in class could not account for its running low so quickly. Rachel and Dorothy suspected that some students might be using the alcohol for nonacademic purposes, especially since drinking alcohol was prohibited during the era of Prohibition. Rachel put a drop of red dye in the bottle that turned the liquid pink and then put a label on it with a skull and crossbones. Never again did the alcohol supply run so low. During her senior year, Rachel was goalkeeper for the field hockey team, which was nicknamed "Navy." She somehow managed to procure a goat, the traditional Navy mascot, as well as its kid for the big championship game, causing quite a humorous commotion. Rachel's team won, and she never revealed where she got the goat.

Going home for the summer was always difficult for Rachel. After the freedom and intellectual stimulation of college life, she found it hard to adjust to living with her family in their cramped house. The Carsons' small two-story home consisted of just four rooms and had neither central heating nor indoor plumbing. Her sister Marian had divorced her second husband, and she was living there with her two daughters. During the summer of 1926, Rachel's brother Robert and his wife and baby daughter had to live in a tent in the backyard. Rachel felt that she had no time or place to herself because she was expected to help out with the children. As much as possible, she tried to find solace as she had always done, walking in the woods and communing with nature.

## SWITCHING HER MAJOR TO BIOLOGY

When Rachel was a sophomore, she took a class that would change her life forever. To fulfill a science requirement, she signed up, like most PCW students, for biology. The biology seminar was taught by thirty-five-year-old Professor Mary Scott Skinker, a passionate and inspiring teacher who had very high standards. Rachel soon discovered that biology captivated her even more than literature as a way of appreciating nature. As Linda Lear explained, "Biology revealed yet another way for Rachel to love nature. Her cognitive and observational skills were suited to it in the same way that her poetic skills enabled her to transcribe what she saw outdoors. Biology did not replace her love of observing nature or writing about it. Rather it reinforced her passion for the mystery and meaning of life" (Lear 1997: 39). During that era, however, the arts and sciences were considered disparate disciplines, so Rachel struggled to choose between them, not realizing that they could be complementary perspectives for looking at the world. She contemplated changing her major to biology but was concerned about losing the school's financial support.

One stormy winter night during her sophomore year, when Rachel was struggling to make a decision about her course of study, she was reading the poem "Locksley Hall" by Alfred, Lord Tennyson. When she read the line, "For the mighty wind arises, roaring seaward, and I go," she felt that Tennyson was speaking directly to her. She later wrote to her friend Dorothy Freeman, "I can still remember my intense emotional response as that line spoke to something within me seeming to tell me that my own path led to the sea—which then I had never seen—and that my own destiny was somehow linked with the sea" (quoted in Freeman 1995: 59).

During this time, Rachel wrote a story called "Broken Lamps," which reflected her inner turmoil over the decision to change her major. In it, an engineer wants to build the perfect bridge that unites strength and beauty, but after his wife's sudden illness, he realizes that the perfect combination of strength and beauty is in his wife, not a physical structure that he can ever build. This story won the annual prize offered by Omega, the literary club, and was published in the *Englicode* on Rachel's twentieth birthday, May 27, 1927.

In the beginning of her junior year, deciding to continue as an English major while minoring in biology, Rachel felt that she had resolved her situation. She enjoyed her classes including a novel course, composition, and vertebrate biology and hygiene. But as the semester progressed, she felt more strongly that Professor Skinker rather than Professor Croff was her true mentor and role model. In January 1928, with only three semesters until graduation, Rachel officially changed her major to biology. She replaced her novel course with a chemistry course and had to devote herself to making up all the labs and courses required to fulfill the major in time for graduation. The administration of the college was disappointed with her decision to change her major because she had held so much promise as a writer. Science was not considered a woman's field in those days, and there were few opportunities for women aside from teaching. But a relieved Rachel was confident that she had made the right choice.

Rachel's only regret was that she had apparently abandoned her writing pursuits. Although popular science writing, the type of writing at which Carson would truly excel, was not very common at the time, she eventually found a large and appreciative audience for her work. She later wrote, "I had given up writing forever, I thought. It never occurred to me that I was merely getting something to write about" (quoted in Lear 1998: 149). She told Dorothy Thompson, "I have always wanted to write, but I know I don't have much imagination. Biology has given me something to write about. I will try in my writing to make animals in the woods and waters where they live as alive and as meaningful to others as they are to me" (quoted in Brooks 1980b: 138). Paul Brooks later wrote, "The merging of these two powerful currents—the imagination and insight of a creative writer with a scientist's passion for fact—goes far to explain the blend of beauty and authority that was to make her books unique" (Brooks 1972: 18).

During Rachel's senior year, Mary Scott Skinker had taken a leave of absence from PCW to pursue her doctorate in zoology. Skinker had never gotten along very well with President Coolidge, who blamed her for Rachel's change of major. Skinker's departure greatly upset Rachel, who regarded Skinker as both her mentor and her friend. Rachel even considered leaving PCW and in April 1928 applied for graduate standing in zoology at Johns Hopkins University, where Skinker was pursuing her doctoral studies. She was accepted but was unable to attend for financial reasons.

Rachel and her friends Mary Frye and Dorothy Thompson disagreed with President Coolidge over the direction and focus of the science departments at the school after Skinker's departure. A new professor hired by Coolidge as a temporary replacement, Dr. Anna Whiting, shared the president's views that the students should ultimately be preparing for marriage and motherhood. The serious science majors who planned on pursuing scientific careers after graduation did not consider Whiting a very good lab instructor.

Rachel had to take six science courses during her senior year to fulfill graduation requirements. To prove to President Coolidge and Dean Marks that her decision had been the right one, she had to excel in all of these classes. In addition to taking her science requirements, she also took a year of German, proofread the *Arrow,* and continued to be involved in athletics. Rachel became president of the science society that she had formed with Mary and Dorothy. They named it Mu Sigma Sigma (MSS) in honor of Mary Scott Skinker.

In the fall of 1928, Rachel again applied to Johns Hopkins University graduate school. The following April, she learned that she had been awarded a full tuition scholarship of $200, which was essential because she had amassed $1,600 in debt from her undergraduate education. She had also won a place at the Marine Biological Laboratory (MBL) at Woods Hole for the summer as a "beginning investigator" or intern after Skinker nominated her. The MBL was a renowned center for the study of marine biology located on the southwestern tip of Cape Cod. Rachel's elation was dampened by the news that Skinker had been unable to pursue her studies at Johns Hopkins because of health problems and had officially resigned from PCW. On June 10, 1929, Rachel graduated magna cum laude, one of only three to do so in her class of seventy students. The editors of the *Arrow* included "Rachel's brains" in their annual description of the "Ideal College Senior."

Before Rachel began her trip to Cape Cod, she returned home to Springdale. By this time, farms had been replaced by smokestacks, the river had been polluted by industrial waste, and the smell of sulfur, a by-product of the coal industry, rose in the air. Rachel's hometown, where she had first learned to appreciate nature, had become a dirty and ugly place that was easier for her to leave behind.

# Chapter 2

# SEAWARD SHE GOES

At the age of twenty-two, Rachel Carson, who had never been more than sixteen miles from her home, left the landlocked state of Pennsylvania to see the ocean that she had dreamt about for most of her life. In the summer of 1929, before going to Woods Hole, she stopped at Johns Hopkins University in Baltimore, Maryland, to register for graduate school classes and visited Mary Scott Skinker in Virginia. She then took a train to New York City where, on a rainy evening, she boarded a passenger boat to New Bedford, Massachusetts. Despite the weather, Rachel lingered on deck as the ship passed the Statue of Liberty and headed out of the harbor to the open sea. In the morning, she took another boat that made the short trip from New Bedford to Woods Hole up Buzzards Bay. Even amidst rough waters, Rachel preferred staying above on deck, thrilled that she was actually on an oceangoing vessel.

## FIRST SUMMER AT WOODS HOLE

Woods Hole, a village belonging to the town of Falmouth, Massachusetts, was located on a peninsula situated on the southeastern tip of Cape Cod. In the middle of the nineteenth century, both professional and amateur naturalists started coming to Woods Hole to study the wide variety of aquatic specimens that could be found offshore. The headquarters of the U.S. Fish Commission, which later became the Bureau of Fisheries, had been located at Woods Hole since it was established as the first federal conservation agency in 1871. Woods Hole also became home to the

Woods Hole Oceanographic Institute, founded in 1930, the year after Carson's first visit.

The Marine Biological Laboratory (MBL) in Woods Hole was founded in 1888 by the Women's Education Association of Boston with the assistance of Alpheus Hyatt, the curator of the Boston Society of Natural History. The MBL was intended to be a place where the study of basic life processes common to all creatures would be facilitated utilizing marine organisms. Because it was an independent research facility, scientists had more freedom than they generally had in the academic world or government agencies. Whereas the Fish Commission was an all-male domain, the MBL began as a place where both men and women could conduct research, and women were encouraged to be a part of the community of scientists there. Because lab tables were assigned based on research project rather than gender, women were not segregated from their male colleagues. Neither did level of knowledge or experience separate those at the MBL, where students and professors worked together, eliminating the hierarchy that characterized other institutions.

Of the seventy-one beginning investigators at Woods Hole during the summer of 1929, thirty-one were women, a very high number compared with most other scientific institutions. Even though Rachel Carson was only an intern, she found herself working alongside renowned scientists and Nobel Prize winners. She dedicated herself to lab work at the MBL and spent hours reading rare books and technical scientific journals in the library. She also took advantage of the ideal natural location to engage in field research, spending time on the MBL dredging boat that would pull up fascinating creatures out of Buzzards Bay and Vineyard Sound. One of her most memorable experiences was spending a day on the U.S. Bureau of Fisheries' research ship, the *Albatross II,* to collect specimens.

Rachel's main project that summer was studying the cranial nerves rooted in the brains of reptiles. She had explored this subject in college and hoped to expand her research for her master's thesis. Specifically, she worked on a comparative study of the terminal nerves in lizards, snakes, and crocodiles, a project that was suggested to her by her academic advisor at Johns Hopkins, Dr. Rheinart P. Cowles, who was also at the MBL that summer. The purpose of the terminal nerves, a pair of cranial nerves that extend from the front of the brain to the nasal area, was not known. While at Woods Hole, Rachel also revised dissection directions for the cranial nerves of turtles, which Cowles thought good enough to be published. She felt, however, that her skills were inadequate to perform her research because of the inferior training she had received from Professor Anna Whiting, Mary Skinker's replacement at Pennsylvania College for Women.

Although the MBL scientists were industrious, their institution, located in a popular vacation area and open only during the summer, exuded a vacation-like quality and was considered "a biological resort, oriented, like other American resorts, around community, relaxation, and nature" (Pauley 1988: 128). Rachel found time to roam the beaches, pursue such recreational activities as tennis and swimming, and even attend an occasional beach party. Mary Frye, Rachel's college friend, was also an intern that summer, and they roomed together in an apartment house across the street from Rachel's lab, where they enjoyed the luxury of hot and cold running water. At the MBL Mess Hall, they felt exceedingly pampered, served by white-coated waiters and eating at linen-covered tables.

Rachel loved everything about Woods Hole, and after her experience at the Marine Biological Laboratory, she was more confident than ever that her choice of a career was the right one and that her decision to change her major had not been in vain. As she engaged in stimulating research with the other students and scientists, she felt a strong sense of belonging. Her experience at the MBL was "both a rewarding intellectual experience and an intensely spiritual time" that "reaffirmed her decision for science" (Lear 1997: 62).

The short time that Rachel spent at Woods Hole had a profound effect on her future writing. In an author's questionnaire that she completed for Simon & Schuster, the publisher of her first book, *Under the Sea-Wind*, she later wrote that at Woods Hole she "began to get my first real understanding of the real sea world—that is, the world as it is known by shore birds and fishes and beach crabs and all the other creatures that live in the sea or along its edge." She surmised, "Probably that was when I first began to let my imagination go down through the water and piece together bits of scientific fact until I could see the whole life of those creatures as they lived them in that strange sea world" (quoted in Lear 1998: 54). It was at Woods Hole that Rachel "first became really aware of the unseen ocean currents" as she put it, continuing, "I am sure that the genesis of *The Sea Around Us* [her second book] belongs to that first year at Woods Hole, when I began storing away facts about the sea—facts discovered in scientific literature or by personal observation and experience" (quoted in Brooks 1972: 111).

## GRADUATE STUDIES AT JOHNS HOPKINS

Rachel began graduate studies at Johns Hopkins University in the fall of 1929 and received a $200 stipend that almost covered the tuition. She studied genetics and was taught by two renowned teachers, Herbert

Spencer Jennings, professor of experimental zoology, and geneticist Raymond Pearl. She did her thesis research under her advisor, Dr. Cowles, who had been at the MBL the previous summer.

Soon after starting at Johns Hopkins, Rachel met with an acquaintance of Mary Scott Skinker, biologist Elmer Higgins, who was the acting director of the Division of Scientific Inquiry, a part of the U.S. Bureau of Fisheries. Rachel wanted to discuss job opportunities and find out how to best prepare for a career as a marine biologist. Having conducted research with scientists from the Bureau of Fisheries at the MBL during the summer, she was interested in learning more about the function of this department and what was being planned for the future. She particularly wanted to know how a woman would fit into this organization.

Although Higgins considered Carson to be knowledgeable and articulate, he told her that women would find few opportunities to work in the sciences except in government agencies or academic institutions. Finding a job in industry or even a museum was rare for a woman. Female scientists were not considered suited to the outdoor fieldwork that such positions might require. Despite Higgins's comments, Rachel's enthusiasm for marine biology was undiminished, and she approached her graduate studies with great determination.

The atmosphere in graduate school was much more intense and competitive than at Woods Hole. Rachel formed no friendships with the other students who were in her program, but she impressed them with her diligence and aptitude. Rachel had a full-time course load during her first year that included Botany 1, a genetics laboratory, and organic chemistry, a difficult class in which she was one of only two women among more than seventy students. Her classes and lab work, which kept her occupied for forty-six hours a week, were supplemented by countless hours of reading and homework.

Just weeks after Rachel started her graduate program, the stock market crashed and the country entered the Great Depression, a period that lasted throughout the decade of the 1930s. Rachel's family had never prospered financially, and the Depression worsened their situation. Mr. Carson's health was deteriorating, and his business pursuits had not lived up to expectations. Rachel had not seen or talked to her family in months because she could not afford to visit or even telephone, and she particularly missed the emotional support of her mother. Lonely and tired of renting a room, Rachel thought that the Baltimore area might hold more opportunity for her family than Springdale. So she persuaded them to rent their home and move in with her, a common arrangement for families during that time.

Rachel rented a house on the Old Philadelphia Road at Stemmers Run, thirteen miles from Baltimore, but just two miles from Chesapeake Bay. Her parents moved there in the spring of 1930, followed by her divorced sister Marian and two young nieces Virginia and Marjorie. Finally, brother Robert, who had separated from his wife in 1927, moved in during the spring of 1931 and took a job in Baltimore repairing radios. One day he came home with a family of Persian cats that was partial payment for a repair job. The mother cat and three kittens completed the Carson family and became cherished pets.

During the summer of 1930, after her first year of graduate studies, Rachel began to teach summer school at Johns Hopkins. Although she had wanted to return to Woods Hole, she needed to earn money to pay her tuition. She also wanted to gain some practical teaching experience, which would be essential if she was to pursue a career in the academic world. Rachel was a teaching assistant in an undergraduate general biology course taught by Grace Lippy, who had earned her master's degree in 1926 under Dr. Cowles and was the only female instructor of zoology at Johns Hopkins at that time. The two women got along well, Grace handling lectures and Rachel the laboratory work. Among Rachel's responsibilities was preparing the lab for forty-five students and making sure that they had all the equipment they needed for their experiments. Grace and Rachel taught together for the next four summers, and Grace became Rachel's only friend at Johns Hopkins.

Financial problems prevented Rachel from taking classes full-time during her second year of graduate work. Although she again received a $200 stipend, the tuition had risen to $300, an amount she could not afford. Unfortunately, she had to decline the scholarship, become a part-time student, and get a part-time job. She worked for a year as a laboratory assistant for Raymond Pearl at the Institute for Biological Research in the School of Hygiene and Public Health at Johns Hopkins, where she was impressed by the number of women on the staff. When this assistantship ended, she was able to locate a part-time teaching job with the help of Dr. Cowles, and in September 1931 she became the only female biology instructor at the Dental and Pharmacy School at the University of Maryland.

## COMPLETING HER MASTER'S THESIS

The 1930–31 academic year was a challenging time for Rachel, both financially and academically. She had to take two laboratory courses in genetics and physiology and was having difficulty honing in on a research topic for her thesis. She was making no progress with her research on the cranial nerves of reptiles because her experiments with reptile embryos had been unsuccessful. By the summer of 1931, Rachel was done with the coursework for her master's degree and only had to complete her thesis to graduate.

Feeling a bit anxious about her master's project, Rachel asked Dr. Cowles for advice. He suggested studying the pronephros, or temporary kidney, of the channel catfish. The pronephros starts to develop just two days after fertilization of the egg. After the catfish has become a larva by the sixth day, a permanent kidney, the mesonephros, starts to develop. The pronephros turns into the head kidney, so called because it lies toward the front of the body. Scientists were uncertain of the function of the head kidney.

Rachel's research would focus on the growth of the pronephros from its beginning in the egg through its eleventh day in the larva, but the ultimate function of the head kidney would remain a mystery. Her project required her to examine hundreds of catfish eggs and larvae under a microscope and read thousands of pages of technical material about catfish and fish kidneys. She painstakingly prepared numerous drawings of her specimens using a camera lucida, a device that uses prisms and mirrors to project an object's image onto a flat surface so that it can be traced. She also included some photomicrographs—photographs made through a microscope.

Rachel finally completed her 108-page thesis, "The Development of the Pronephros during the Embryonic and Early Larval Life of the Catfish (*Ictalurus punctatus*)" in the spring of 1932. She was awarded her master's degree in marine zoology on June 14, 1932. About her experiences at Johns Hopkins, she later recalled, "Whatever else I may have learned there, this was the unforgettable lesson: we do not really know anything. What we think we know today is replaced by something else tomorrow" (quoted in Brooks 1972: 206). After Rachel finished teaching with Grace Lippy during the summer of 1932, the two of them went to Woods Hole and roomed together. As an "investigator," Rachel worked in the Bureau of Fisheries lab, but it is not known what she studied there that summer.

## HARD TIMES DURING THE DEPRESSION

Rachel wanted to continue graduate work for her doctorate in preparation for a teaching career and had even registered for classes. Her financial situation and family responsibilities, however, made this impossible. Rachel was unable to start repaying her debt to the Pennsylvania College for Women in 1931 as required. Eventually she had to give up the two lots of the family property that she had used as collateral. She dropped out of the doctoral program officially before the spring semester of 1934.

Rachel needed to find a full-time job, a rare opportunity during those Depression years when one in every four workers was unemployed, and harder still because she was a woman trying to enter a male-dominated field. Opportunities for marine biologists, whether male or female, were practically nonexistent at that time. She looked for a full-time teaching position at four-year and community colleges rather than universities, because of her lack of a Ph.D. Without a doctorate, any advancement in the academic world would be almost impossible. Rachel taught part-time at the University of Maryland during the academic year until the end of 1933. Over most of the next two years, her only source of income was the money she earned teaching at Johns Hopkins' summer school.

Rachel's family situation was complicated by illness. Her sister Marian had become diabetic and could not work on a regular basis, and her father was becoming increasingly frail. So Rachel alone had to support financially three other adults as well as two children. With no full-time employment prospects, she decided to try making some money from writing. She hadn't written much since college, so she found the best of her poems and short stories from that time, revised them, and sent them to a variety of national magazines including the *Saturday Evening Post* and *Reader's Digest*. She received nothing but rejection slips, just as she had when she had sent them out during college. Despite this negative reaction to her work, she rediscovered her love of writing through this process.

Rachel stayed in touch with Mary Scott Skinker, who served as her "model, both of what a scientist was and what she herself might expect from a life spent in the practice of that discipline" (Lear 1997: 57). Skinker worked as a government scientist at the Zoological Division of the Department of Agriculture's Bureau of Animal Industry, having completed her Ph.D. in zoology at George Washington University in 1933. She advised Rachel to take the civil service exams in zoology so that, in the event that a position might open up in a government agency, Rachel would be able to apply for it.

In 1935, Rachel took the junior-level exams for parasitologist, wildlife biologist, and aquatic biologist and attained outstanding scores. Although a number of sources claim that she scored highest on the aquatic biologist test, at that time separate registers were kept for men and women by the Civil Service Commission. She placed first on the women's register, but it is not certain whether she was the highest scoring applicant. Rachel was encouraged by her performance on the exams, but still had no job prospects.

As the Depression wore on, President Franklin D. Roosevelt offered re-assurance to many people. Interested in conservation as his distant cousin Theodore Roosevelt had been, he directed his government to create many jobs involving public works. At the same time, First Lady Eleanor Roosevelt, who advocated the rights of women, helped change deep-seated attitudes about gender roles. As more opportunities opened up to women, Rachel gained new hope for her own future.

Rachel's optimism was dampened when her father died suddenly on July 6, 1935, at the age of seventy-one. His body was sent back to Canonsburg, Pennsylvania, where his sisters buried him; Rachel and her mother and siblings could not afford to attend. As Rachel became her mother's sole financial supporter, she also had to help her ailing sister and two young nieces even more.

## WRITING FOR THE BUREAU OF FISHERIES

As Rachel's persistent search for employment gained a new urgency, she again visited Elmer Higgins at the Bureau of Fisheries. Although he had no permanent full-time job openings, he was in need of a writer on a temporary part-time basis. Scripts needed to be written for "Romance under the Waters," a public education series of fifty-two short radio pro-grams on marine life, nicknamed "seven-minute fish tales."

Higgins realized that the staff scientists couldn't write in an engaging manner, and other professional writers didn't have the necessary scientific background. When Rachel told Higgins that she had started out as an En-glish major in college, he hired her on the spot to write the scripts, and she became a part-time field aide at the Bureau of Fisheries. The scripts were well received, and Rachel was pleased to earn $19.25 per week.

Rachel also started writing feature articles for the local newspaper, the *Baltimore Sun*, using some of her research from the radio scripts. Her first article, "It'll Be Shad-Time Soon—and Chesapeake Bay Fisherman Hope for Better Luck This Season," appeared in March 1936. This piece was written under the byline of R. L. Carson because Rachel thought she would be taken more seriously if readers assumed she was a man. At a time when the environment was hardly a concern of the average American, this article exposed the pollution of the Chesapeake Bay that threatened the survival of the fish that lived in it. Rachel published numerous articles for the *Sun* for which she received $10 to $20 apiece.

After Rachel had finished the radio scripts for the Bureau of Fisheries, Higgins asked her to continue writing for the department part-time. In April 1936, Rachel wrote an essay, "The World of Waters," for a general information brochure on marine animals. Higgins was impressed and thought it was much too good to be published in a government brochure. Because he considered the essay to be a piece of literature, he suggested that Rachel submit it to the *Atlantic Monthly,* one of the premier literary magazines in the country. Instead, Rachel, who did not share Higgins's level of enthusiasm, sent it to a contest sponsored by *Reader's Digest* and awaited a reply.

## JUNIOR AQUATIC BIOLOGIST

Meanwhile, an entry-level position as a junior aquatic biologist with the Division of Scientific Inquiry opened up in July 1936. Although Rachel, with her graduate degree, was overqualified, she applied and was appointed because of her high score on the civil service exam. On August 17, 1936, at age twenty-nine, Carson began working full-time at the Bureau of Fisheries at an annual salary of $2,000. As a junior aquatic biologist, Carson's duties included writing and editing various publications, such as reports and brochures for a general audience about fish conservation, and answering many questions from the public concerning fish. She also aided Robert Nesbit, the assistant bureau chief, with his study of Chesapeake Bay fish and analyzed the data collected by a group of scientists to determine age and population statistics. She got out of her cramped office from time to time to locate information in libraries and to visit scientists in other laboratories and field stations as part of her research.

Carson's boss, Elmer Higgins, became another mentor, providing a constant source of support and encouragement for her writing endeavors. He was grateful to have someone with her knowledge and writing ability working at the department. Rachel felt fortunate to have such a supervisor because she experienced much less of the discrimination that other women scientists faced at that time. According to Philip Sterling, Higgins "had given her a clue to harmonious living with her twin loves, art and science," and had helped her to realize that "she had something everlastingly worthwhile to write about—something as rich in real-life beauty as anything that poets could imagine" (Sterling 1970: 94).

Tragically, in January 1937, Rachel's sister Marian died of pneumonia at age forty. This event forced Rachel to give up what little freedom and independence she had retained after the death of her father had increased her family responsibilities. In addition to supporting her mother, she now had to take care of raising nieces Virginia, twelve, and Marjorie, eleven, who had lost their mother. She also gave up her proximity to Chesapeake

Bay, moving from Stemmers Run to a larger house in Silver Spring, Maryland, where she could more easily commute to her job in Washington, D.C., and her young nieces could attend a better school.

Although Rachel had a secure job, a steady income, and a promising career as a marine biologist, she considered her salary to be inadequate for her new financial burdens. She was grateful to have a job that was suited to her talents and work that combined her interests in biology and writing, but she felt confined in an office all day long. As she continued to dream of being a full-time writer, she wrote at night and on weekends. By June 1937, having not received a reply from *Reader's Digest* regarding "The World of Waters," she followed Higgins's advice and submitted a revised version to the *Atlantic Monthly*, an action that would ultimately change the course of her life.

# Chapter 3

# MAKING THE SEA A "VIVID REALITY"

## "UNDERSEA"

On July 8, 1937, the acting editor of the *Atlantic Monthly*, Edward Weeks, wrote to Rachel Carson, "We have everyone of us been impressed by your uncommonly eloquent little essay, 'The World of Waters.' The findings of science you have illuminated in such a way as to fire the imagination of the layman" (quoted in Lear 1997: 86–87). Weeks suggested, however, that Carson change the title of the article to "Undersea," which she did. She received $100 in early August, and the essay was published in the September 1937 issue.

In "Undersea," Carson wrote from the perspective of the creatures that lived there, not as an alien human being. Her remarkably concise four-page article described tide pools, the surface waters of the ocean, the environment just below the surface, and finally, the depths of the sea. She began by acknowledging the inability of humans, portrayed as predators from beyond the ocean world, to truly understand the marine environment, writing:

> Who has known the ocean? Neither you nor I, with our earth-bound senses, know the foam and surge of the tide that beats over the crab hiding under the seaweed of his tide-pool home; or the lilt of the long, slow swells of mid-ocean, where shoals of wandering fish prey and are preyed upon, and the dolphin breaks the waves to breathe the upper atmosphere. Nor can we know the vicissitudes of life on the ocean floor, where the sun-

light, filtering through a hundred feet of water, makes but a fleeting, bluish twilight. . . . Even less is it given to man to descend those six incomprehensible miles into the recesses of the abyss. (quoted in Lear 1998: 4)

"Undersea," which explained the interrelationships among sea creatures and their environment, reflected Carson's ecological belief that all living things are related to each other. After describing the ocean floor as covered with the remains of dead plants and animals that had fallen from above, she wrote, "Every living thing of the ocean, plant and animal alike, returns to the water at the end of its own life span the materials that had been temporarily assembled to form its body" (quoted in Lear 1998: 9). Thus, organisms that die provide material that can be of use to the living. Carson explained the intricate food chain that links microscopic diatoms and plankton to larger creatures such as the great blue whales and revealed how all creatures are dependent on each other in the marine ecosystem. Big fish hunt for smaller creatures and are themselves preyed on by larger organisms.

Carson's *Atlantic Monthly* article "established her unique voice, at once scientifically accurate and clear, yet with poetic insight and imagination, one that confidently captured the wonder of nature's eternal cycles, rhythms, and relationships" (Lear 1997: 88). As she attempted to describe the deepest part of the ocean, she asked in her characteristic rhetorical fashion, "What human mind can visualize conditions in the uttermost depths of the ocean?" (quoted in Lear 1998: 8). Apparently *her* mind could accomplish this task as she went on to describe in vivid detail this alien environment. She enabled the reader not only to see but also to feel the intense cold, darkness, and crushing pressure that is experienced by the strange creatures of the sea. "Undersea" possessed a poetic style that communicated complicated scientific material in a way that could be understood by the general reader, a style that would become Carson's trademark. Rachel said that from this short article "everything else followed" (quoted in Lear 1998: 150).

## SPARE TIME WRITING PURSUITS

Quincy Howe, senior editor at publisher Simon & Schuster, read "Undersea" and encouraged Rachel to write an entire book on the same subject. Howe edited the books of the well-known historian Hendrik Willem van Loon, author of *The Story of Mankind* (published in 1921), who also admired Carson's writing and urged her to develop her ideas into a book. Because van Loon frequently crossed the ocean by boat, he had become very curious about what lay beneath the seemingly lifeless sea.

Carson met with Howe at van Loon's home in Old Greenwich, Connecticut, in January 1938, and with their help and encouragement, she developed a rough outline. She envisioned a book in which sea creatures would be the central characters and humans would only be present on the periphery. After completing one chapter and sending that along with the outline to Simon & Schuster, she received only a $250 advance. She would not be offered a contract until the editors saw a few more chapters. Because the advance was so small, she was unable to devote all of her free time to the book. She needed money and continued writing articles for publications that would pay her immediately.

During this time, Carson wrote a series of well-researched feature articles for the *Baltimore Sun* focusing on fish and the fishing industry in the Chesapeake Bay area. One article, "Chesapeake Eels Seek the Sargasso Sea," which appeared in the newspaper on October 9, 1938, is a foretaste of a subject to which she would devote a whole section of her book *Under the Sea-Wind*. Because of her lifelong interest in ornithology, she also wrote an article about starlings for the *Baltimore Sun* and later sold a revised version of this article to *Nature Magazine*. "How about Citizenship Papers for the Starling?" was the first piece she wrote under the name of Rachel L. Carson instead of R. L. Carson, perhaps reasoning that she had established herself as an expert and could now reveal that she was indeed a woman. She also sold a few articles to the *Richmond Times Dispatch* on similar topics about the Chesapeake Bay and its creatures and contributed a number of book reviews to the *Atlantic Monthly*.

Rachel was concerned about protecting the environment at a time when the rest of the world seemed ignorant of this problem. She had suggested a couple of topics to the editor of the *Sun*, Mark Watson, concerning pollution, but he thought that readers would not be interested in this subject. In an article for the *Richmond Times Dispatch* titled "Fight for Wildlife Pushes Ahead," which reflects her ecological perspective, she wrote, "The inescapable fact that the decline of wildlife is linked with human destinies is being driven home by conservationists the nation over. Wildlife, it is pointed out, is dwindling because its home is being destroyed. But the home of wildlife is also our home" (quoted in Lear 1998: 15). She stressed that humans and animals share an environment and that decreases in wildlife populations should certainly be of concern to everyone.

For three years, at night and on weekends, Rachel worked as much as possible on the book that was eventually titled *Under the Sea-Wind: A Naturalist's Picture of Ocean Life*. She wrote out her words in longhand at night while her cats kept her company during those lonely hours, and her mother typed them up during the day while Rachel was at work. Although a few of her coworkers knew she was writing something substantial outside of the office, she remained secretive about the details of her project.

In the fall of 1939, Rachel and her family moved to a quieter section of Silver Spring to a house where Rachel could have some privacy in a bedroom that occupied the entire second floor. By the spring of 1940, she had completed the first five chapters of *Under the Sea-Wind* and hoped that this would be enough for Simon & Schuster to offer her an official contract. She finally received a contract in June 1940 that required her to complete her book by the end of the year. Work on the book accelerated and became her priority once she had a contract, a deadline, and an additional advance of $250.

In the midst of working on the book, Rachel returned to Woods Hole during the summers of 1939 and 1940. She did some research in the library and some laboratory work, but mostly these trips gave her the opportunity to examine some of the creatures about which she was writing. While she was on Cape Cod in 1940, Rachel and her colleague and friend Dorothy Algire from the Bureau of Fisheries went to Eastham, Massachusetts, where they visited the small cottage where writer Henry Beston had lived while writing *The Outermost House*, published in 1924. This book, one of Rachel's favorites, is an account of the year Beston spent alone on the isolated beach. Rachel, who was strongly influenced by the work of Beston, wrote to a friend, "It is written with great simplicity and beauty, and with a feeling for the great rhythms of nature" (quoted in Brooks 1972: 5). Her literary pilgrimage to Beston's cottage inspired her, and her time at Woods Hole enabled her to finish her book.

In November 1940, Carson mailed her manuscript, carefully typed by her mother, to Simon & Schuster. *Under the Sea-Wind* was published in November 1941, one month before the attack on Pearl Harbor that led the United States into World War II. Only 1,348 copies were sold during its first year in publication, and a total of 1,600 over six years in print. The public seemed to be preoccupied with the war at that time and not interested in reading such a book. As Carson herself put it, "The world received the event with superb indifference" (quoted in Lear 1998: 150). Although the war might be partly to blame for poor sales, she also felt that the book did not sell well because Simon & Schuster did not promote it effectively.

## AN OVERVIEW OF *UNDER THE SEA-WIND*

In the preface to the first edition of *Under the Sea-Wind*, Carson explained that her reason for writing the book was "to make the sea and its life as vivid a reality for those who may read the book as it has become for me during the past decade." She went on to write that the book was written "out of a deep conviction that the life of the sea is worth knowing" (quoted in Brooks 1972: 32). She summarized her work as "a series of descriptive narratives unfolding successively the life of the shore, the open sea, and the sea bottom" (34). The book is divided into three parts: "Edge of the Sea," which follows the migration of sea birds; "The Gull's Way," about a mackerel who escapes a number of close calls with death; and "River and Sea," featuring an eel returning to its birthplace. Carson felt that the three parts of the book taken together would reflect the interdependence of all the ocean's creatures.

The first part of *Under the Sea-Wind*, "Edge of the Sea," features Blackfoot, a type of sandpiper known as a sanderling. Blackfoot is the leader of the flock, and his mate, Silverbar, is a yearling who is returning for the first time to her birthplace. The flock stops on an island off the coast of North Carolina in the process of migrating from its South American winter home to its nesting place in the Arctic tundra. Another character in this first section is a black skimmer named Rynchops, a member of a flock that has flown to the island to lay and hatch its eggs before returning to the Yucatan. Carson chose these birds as the main characters because their long-range migration habits fascinated her. The setting for the beginning island episode of this section was based on a portion of the Outer Banks near the U.S. Fisheries Station in Beaufort, North Carolina, that Carson first visited for ten days in July 1938. She returned there many times and spent countless hours during the day and at night observing shore creatures at high and low tide.

When the flock arrives at its northern destination, the birds struggle for survival during a bad snowstorm. In the Arctic nesting place, another character, Ookpik the snowy owl, must abandon her nest and eggs to survive. Because of the resultant death of six "owls-to-be," Carson claimed that "hundreds of unborn lemmings and ptarmigans and Arctic hares had the greater chance of escaping death from the feathered ones that strike from the sky" (Carson 1941: 52). This episode demonstrates the delicate balance of nature and how all species are intricately interconnected.

The second part of *Under the Sea-Wind* features a mackerel named Scomber. The reader follows him from the beginning of his life as a tiny egg drifting on the ocean surface until he joins a school of mackerel in a New England harbor and goes with them to the open sea. It is another story of long-distance migration—this time by a fish rather than a bird. The dramatic narrative structure of the entire book is particularly evident here as Scomber repeatedly has close calls with death, often escaping because those who hunt him are preyed on by other creatures. As in the other episodes of *Under the Sea-Wind*, a sense of sympathy is elicited for the hunted creatures of the sea, and readers are happy when Scomber ultimately survives, later in Book 2, after some dogfish with sharp teeth cut holes in the fishing net in which he is trapped.

In the third part of *Under the Sea-Wind*, the ocean world is seen from the perspective of Anguilla the eel, who leaves the inland pond where she has spent ten years of her life. She travels two hundred miles back to the spot in the ocean abyss where she was born in order to give birth to a new generation of eels. Eels from both America and Europe migrate to this same distant spot, south of Bermuda in the Sargasso Sea, but remarkably, the offspring always return to the homeland of their parents whom they have never known. Carson's descriptions of this environment far below the surface of the sea are particularly interesting because this is an area so alien to human beings and so unlike the shore or the open sea.

*Under the Sea-Wind* concludes with an extensive glossary that adds scientific accuracy to Carson's work. It contains concise yet literate descriptions of all the creatures featured in the narrative. Howard French, staff artist for the *Baltimore Sun*, provided the realistic illustrations included in the glossary and throughout the book.

With a perspective similar to that of her *Atlantic* article, in *Under the Sea-Wind* "Rachel had succeeded in thinking her way aloft and underwater to picture the ocean world from the point of view of its inhabitants" (Sterling 1970: 100). The stories are told from the perspective of the sea creatures themselves as they face their daily struggles for survival. Earlier, in 1938, Carson had written to Hendrik Willem van Loon, "The fish and the other creatures must be the central characters and their world must be portrayed as it looks and feels to them." She continued that humans must not "come into it except from the fishes' viewpoint as a predator and destroyer" (quoted in Lear 1997: 90). The entire book is told from the point of view of an omniscient narrator who can get inside the characters of birds, fish, and, on a single occasion, a fisherman. Overall, however, Rachel considered the sea itself to be the central character of the book.

Although human beings generally remain on the periphery in *Under the Sea-Wind*, one episode is told from the viewpoint of a fisherman. He is different from the other fishermen because he has not yet forgotten "the wonder, the unslakable curiosity he had brought to his job—curiosity about what lay under the surface" (Carson 1941: 200). Not merely regarding the mackerel as an object of prey, he sometimes wonders what this creature of the sea has seen and experienced, and even wishes that he could be down in the sea with the fish he is hunting. Carson provided this fisherman's attitude as a model for her readers while she criticized the perspective of the other fishermen who thoughtlessly destroy the ecology of the ocean environment and threaten the balance of nature by overfishing, pursuing fish that are struggling upstream to spawn, and fishing purely for sport.

Just like in her *Atlantic Monthly* article, the sea creatures that Carson wrote about are engaged in the daily struggles for survival between hunter and hunted, prey and predator. They are involved in the drama of daily life, and Carson's action-packed descriptions reflect this natural process that preserves the balance of nature. She revealed the intricate workings of the food chain, writing, "For in the sea nothing is lost. One dies, another lives, as the precious elements of life are passed on and on in endless chains" (Carson 1941: 105). The ecological theme of the book is reflected in the fact that all creatures are eternally interdependent—the death of one leads to life for another. For instance, a ghost crab feeds on beach fleas; he, in turn, is eaten by a channel bass that is then preyed on by a shark.

Throughout the book, Carson anthropomorphized the sea creatures she wrote about, depicting various characters, for example, as "nervous," "cunning," "bickering," and in a "panic"—words that are normally associated with human beings. In the preface to the first edition of *Under the Sea-Wind*, Carson explained her unusual style of scientific writing:

> I have spoken of a fish "fearing" his enemies, for example, not because I suppose a fish experiences fear in the same way that we do, but because I think he behaves as though he were frightened. With the fish, the response is primarily physical; with us, primarily psychological. Yet if the behavior of the fish is to be understandable to us, we must describe it in the words that most properly belong to human psychological states. (quoted in Brooks 1972: 34)

She wanted readers to identify with the characters in order to understand them better, but she also wanted to remain scientifically accurate.

Carson's technique is reminiscent of the English naturalist Henry Williamson, who wrote such books as *Tarka the Otter* and *Salar the Salmon* in which he anthropomorphized animals. Rachel admired Williamson and wrote to her friend Dorothy Freeman in 1953 that her "own style and thought were deeply influenced" by Williamson (quoted in Freeman 1995: 11). She said that his two books would go with her along with Beston's *Outermost House* if she were stranded on a desert island. Rachel's approach is different, however, and more scientifically accurate. She gave names to each of the "characters" based either on the scientific names for the genus the creature belonged to (e.g., Scomber is the scientific name for mackerel) or an appropriate descriptive word (e.g., Blackfoot and Silverbar). Nature writer Richard Jeffries was another strong influence on Carson, especially his *Pageant of Summer*, published in 1905, the theme of which was the sea as the source of life. He wrote, "As the wind, wandering over the sea, takes from each wave an invisible portion, and brings to those on shore the ethereal essence of ocean, so the air lingering among the woods and hedges—green waves and willows—full of fine atoms of summer" (quoted in Lear 1997: 104). This favorite passage of Rachel's inspired the title of her book.

Carson's writing has been praised for its poetic yet factual quality, and in *Under the Sea-Wind* her descriptions of the passage of time, the cycles of the seasons, and the rhythm of nature are both lyrical and scientific. The world she described is ruled by tides and light and darkness. Time, on a human scale, measured by clocks and calendars, is meaningless. She tried to convey this idea in her preface, writing:

> To stand at the edge of the sea, to sense the ebb and the flow of the tides, to feel the breath of a mist moving over a great salt marsh, to watch the flight of shore birds that have swept up and down the surf lines of the continents for untold thousands of years, to see the running of the old eels and the young shad to the sea, is to have knowledge of things that are as nearly eternal as any earthly life can be. (quoted in Brooks 1972: 32)

Another poetic description of the passage of time and the rhythm of nature appears toward the end of the book when the abyss, the deepest part of the sea, is described:

> Below them lay the abyss, the primeval bed of the sea, the deepest of all the Atlantic. The abyss is a place where change comes slow, where the passing of the years has no meaning, nor the swift succession of the seasons. The sun has no power in those depths, and so their blackness is a blackness without end, or beginning, or degree. No beating of tropical sun on the surface miles above can lessen the bleak iciness of those abyssal waters that varies little through summer or winter, through the years that melt into centuries, and the centuries into ages of geologic time. Along the floor of the ocean basins, the currents

are a slow creep of frigid water, deliberate and inexorable as the flow of time itself. (Carson 1941: 261)

Carson later asserted that the subjects she wrote about could "give us a little better perspective on human problems." The stories about Blackfoot, Scomber, Anguilla, and the other characters concern "things that have been going on for countless thousands of years. They are as ageless as sun and rain, or as the sea itself" (quoted in Lear 1998: 62).

## REACTION TO *UNDER THE SEA-WIND*

Although *Under the Sea-Wind* might not have been a popular book, it established Carson's reputation and was praised in both the scientific and literary communities. One critic described the book as being "so skillfully written as to read like fiction, but in fact a scientifically accurate account of life in the ocean and along the ocean shores" (*New York Times* 1941: 27). Another review acclaimed, "Miss Carson's unemotional handling of her subject matter is anything but dull. There is drama in every sentence. She rouses our interest in this ocean world and we want to watch it" (Sutton 1941: 5).

*Under the Sea-Wind* was the November selection of the Scientific Book Club, which asserted in its *Review*, "There is poetry here, but no false sentimentality" (Compton 1941: 1). The book was praised by scientists who generally would dismiss any popularization of science for the general public. Dr. William Beebe, the first person to descend more than half a mile into the ocean in a bathysphere, asserted in a review of the book in the *Saturday Review of Literature*, "Miss Carson's science cannot be questioned; I have been unable to detect a single error" (Beebe 1941: 5). Three years later, Beebe included part of *Under the Sea-Wind* in an anthology that he compiled, *The Book of Naturalists*, along with excerpts from Aristotle, Audubon, and Thoreau.

Despite its failure in the marketplace, *Under the Sea-Wind* was always Carson's favorite work. She felt that she was able to lose herself in the creative process while writing it, putting aside her personal problems and responsibilities in a way that she was unable to do while writing her other books. After the publication of *Under the Sea-Wind*, disappointed by sales, Rachel could hardly imagine that she would ever write another book and decided to limit her freelance writing to magazine articles. She shifted her focus to her job and did what she could to help in the war effort.

# Chapter 4

# FEDERAL EMPLOYEE "IN ACTION"

## MEANWHILE, BACK AT THE OFFICE ...

After the publication of *Under the Sea-Wind*, Rachel Carson continued to excel as a government employee, although she was often frustrated by the bureaucracy surrounding her. Because she was involved in what was considered at that time an acceptable female pursuit—providing public information and serving editorial functions—she was more easily accepted and experienced less discrimination than female scientists working in the field or lab. While she was writing *Under the Sea-Wind*, there were great changes going on in the structure of her organization. In 1940, the Bureau of Fisheries, which was part of the Department of Commerce, had merged with the U.S. Biological Survey, part of the Department of Agriculture, and was renamed the U.S. Fish and Wildlife Service (FWS), an agency of the Department of the Interior. The Secretary of the Interior, Harold L. Ickes, was interested in conservation, so Carson considered this restructuring a positive change.

Carson had begun her full-time career at the Bureau of Fisheries in an entry level "grade P-1" position of junior aquatic biologist. In June 1939, Elmer Higgins recommended her for a promotion to assistant aquatic biologist, a "grade P-2" position offering an annual salary of $2,600 and a relocation from the cramped Baltimore offices to the field laboratory in College Park, Maryland. Due to the reconfiguration of the agency, her promotion was deferred, but she did relocate to College Park. There she continued to assist Robert Nesbit with his research, rewriting all of his laboratory and field reports and compiling bibliographies. She also helped to produce brochures in a series called "Our Aquatic Food Animals." In May 1942, she was finally promoted to assistant aquatic biologist in the Division of Information. In this position, she continued to write a variety of publications about fish, helped Elmer Higgins edit field reports, and edited the *Progressive Fish-Culturalist*, a marine biology journal.

After the war began, the Fish and Wildlife Service was ordered to relocate to Chicago temporarily so that agencies and departments in Washington dealing directly with military matters could have more office space. Rachel, who had already moved several times since coming to the Baltimore area in 1929, was not enthusiastic about uprooting herself to another part of the country, especially considering her family responsibilities. In August 1942, she left with her mother for Chicago, and they rented a small house in nearby Evanston, while her nieces remained behind. Virginia, who worked as a stenographer, and Marjorie, who had recently graduated from high school, stayed with friends.

During this time, Rachel began to feel a certain amount of dissatisfaction with her work and with the federal bureaucracy that employed her because she felt that she was not making a tangible contribution to the war effort. To alleviate her frustration, she qualified as an air raid warden and took a first aid course. When she began to work on an interesting new series of FWS Conservation Bulletins called *Food from the Sea*, she finally felt like she was providing a valuable service. The purpose of the series was to encourage the average citizen to use more seafood because of the rationing of meat and poultry during the war.

The first bulletin that Carson wrote, published in 1943, was titled *Fish and Shellfish of New England,* and the next, also published that year, was *Fishes of the Middle West.* Carson, who later described herself as "only mildly enthusiastic about seafoods," wrote these lengthy booklets in her usual accurate yet engaging style (*Washington Post* 1951: 3B). They were "crammed with facts, yet written in a manner that the American housewife could understand" (Brooks 1972: 72). The bulletins introduced the reader to lesser-known fish because the more popular ones such as cod and haddock were being overconsumed. Each bulletin provided a history of the fisheries in a particular geographic region, descriptions of the fish, and cooking guidelines. Together, the two bulletins that Carson wrote while she was in Chicago and two others that she wrote when she returned to Washington covered almost a hundred species of fish.

As Carson had hoped, her stay in Chicago was brief, lasting less than a year. She applied for a job back in Washington in the Office of the Coordinator of Fisheries as an associate aquatic biologist. Higgins selected her for the "grade P-3" position, giving her a $600 salary increase. She and her mother left Chicago and moved back to Maryland in May 1943 to a small house in Tacoma Park, east of Silver Spring, where they were reunited with Marjorie and Virginia. Rachel began working in the new home of the Department of the Interior—an immense modern seven-story complex that covered two square blocks and was full of spacious offices with large windows and air-conditioning. Aside from additional administrative duties, Carson's job responsibilities remained basically the same: editing field reports, writing press releases and other informational material, and so forth. She wrote two more *Food from the Sea* bulletins: *Fish and Shellfish of the South Atlantic and Gulf Coasts*, published in 1944, and *Fish and Shellfish of the Middle Atlantic Coast*, which came out in 1945.

## INCREASING JOB DISSATISFACTION

Although Rachel was glad to be back home in Maryland, the feelings of job dissatisfaction that she began to experience in Chicago continued, and she was becoming increasingly restless in the governmental bureaucracy that employed her. She felt that she was not truly fulfilling her life's purpose by working at the Fish and Wildlife Service. Her work often seemed tedious and left her little time for herself. As she deprived herself of sleep to find time to write, she began to suffer increasingly from minor health problems. She wrote to Sunnie Bleeker, who worked in the marketing department at Simon & Schuster, "I'm definitely in the mood to make a change of some sort, preferable to something that will give me more time for my own writing" (quoted in Brooks 1972: 75–76).

Rachel wrote magazine articles to supplement her income, believing that this would be the most lucrative type of writing she could do. Sunnie Bleeker assisted in marketing her work, along with Simon & Schuster editor Maria Leiper, both of whom "sympathized with the unyielding financial responsibilities that drove Rachel's efforts to find markets for her writing" (Lear 1997: 106). Her first published article since her book came out three years earlier was "Ocean Wonderland," a piece about the Oceanarium in Marineland, Florida, that appeared in the March 1944 issue of *Transatlantic*. A popular tourist attraction, the Oceanarium was a huge aquarium where thousands of ocean creatures cohabited. Carson's article, which was reprinted in *This Month* in June 1946, explained how the many varieties of fish and mammals were able to live together in an artificial environment that replicated the undersea world.

Carson submitted an article to *Collier's* about bats and their use of radar to fly in the dark. "He Invented Radar—Sixty Million Years Ago!" was based on the research contained in some declassified research reports that

Carson was able to access because of her position at the Fish and Wildlife Service. The article was published in *Collier's* in November 1944 with the new title "The Bat Knew It First." In August 1945, it was reprinted in *Reader's Digest,* which had rejected an article she had submitted about oysters a year earlier. She received $500, an enormous sum to her that amounted to about half of what she had earned for *Under the Sea-Wind.* Because the article so clearly explained the subject, the U.S. Navy decided to reprint it for those recruits who wanted to learn about the technology of radar.

Rachel sent another article to be considered by *Reader's Digest,* "The Ace of Nature's Aviators," about the migration of the chimney swift, a subject that she had been researching at work. After the piece was rejected, she sent it to *Coronet* and was given the opportunity to have it published in condensed form for the small sum of $55. After a five-day stay in a hospital for an appendectomy, she accepted this offer, and the shortened article, retitled "Sky Dwellers," appeared in the magazine in November 1945.

Rachel continued to be frustrated at work and, because her freelance writing would not support her family, began looking for another job. Her frequent rejections from *Reader's Digest* did not deter her from applying for an editorial job with the magazine, but she was told that nothing was open. She contacted Dr. William Beebe about opportunities at the New York Zoological Society, but, despite his fondness for her writing, he was unable to offer her a job. She also applied to be a staff writer at the National Audubon Society for the organization's magazine but was turned down. There were no women in professional positions at any of these organizations, and Carson was looking for a job around the same time that many men were returning from the war in search of work. After her lack of success, she decided to remain at the Fish and Wildlife Service and continued to write on the side, hoping that eventually she could make a living from her freelance writing alone.

Carson's job provided her with access to a lot of valuable information, including research conducted during the war, which helped her in her writing endeavors. Some of this information concerned a new synthetic pesticide known as DDT. Carson became interested in this subject when she edited a research report on the effects of DDT on fish. She thought about writing an article on DDT, specifically on testing being conducted in Patuxent, Maryland, and even queried *Reader's Digest,* but once again the editors were not interested. Neither were any other magazines. So she set this idea aside for the time being and moved on to other projects.

## POSTWAR LIFE

After realizing that her work situation was decent, Rachel tried to enjoy the postwar period. In 1945, she moved back to Silver Spring to a larger house. As usual, her mother took care of the household chores and cooking, leaving Rachel free to write in her spare time or enjoy social activities. Although often characterized as a shy and private person, she liked to socialize among certain people and formed some close friendships at the FWS, making it easier for her to continue working there despite her frustrations. Her supervisor Lionel A. "Bert" Walford, whose book *Marine Game Fishes of the Pacific Coast* she had positively reviewed in the *Atlantic Monthly* in 1938, invited her to many social events with his family. Illustrator Katherine "Kay" Howe and information specialist Shirley Ann Briggs became her good friends and were excellent travel companions on business trips.

Rachel and Bert shared an office and would have lunch there each day with Kay and Shirley. With a shared sense of humor, they tried to bring some fun and laughter to the dull routine of their work. Rachel, who preferred to be called "Ray" by her government friends, was always professional and meticulous in her work and had high standards, but, despite the reserved exterior she often presented to others, she was fun to have around the office. Shirley, who joined the service in 1945, liked working with Rachel because of her sense of humor, which made even the mundane aspects of their work fun. She wrote that Rachel's "qualities of zest and humor made even the dull stretches of bureaucratic procedure a matter for quiet fun, and she could instill a sense of adventure into the editorial routine of a government department" (Briggs 1970: 9).

Outside of work, Rachel enjoyed outings with the local chapter of the National Audubon Society and was elected to its board of directors in 1948. Rachel's social life revolved around her participation in the activities of this group including field trips, morning bird walks in Washington, D.C., and evening lectures. Throughout her life, her interest in birds rivaled her interest in the sea, and she saw similarities between the mountain habitats of some of the birds she enjoyed watching and the ocean environment. In October 1945, she spent two days on Hawk Mountain Sanctuary in eastern Pennsylvania with Shirley Briggs and family friend Alice Mullen. In the notebook she kept during this outing, she wrote, "Perhaps it is not strange that I, who greatly love the sea, should find much in the mountains to remind me of it. I cannot watch the headlong descent of the hill streams without remembering that, though their journey be long, its end is in the sea. And always in these Appalachian high-lands there are reminders of those ancient seas that more than once lay over all this land" (quoted in Lear 1998: 32). On another Audubon trip to Seneca, Maryland, Rachel and Shirley met Louis Halle, a State Depart-ment employee, author, and amateur naturalist whose book, *Spring in Washington*, was published in 1947. Rachel admired his work, and Halle also had a high regard for Rachel, remembering her as "quiet, diffident, neat, proper, and without any affectation," as well as "always attentive, al-ways listening, always wanting to know" (quoted in Brooks 1972: 97–99).

## CONSERVATION IN ACTION

Rachel was rising steadily at the Fish and Wildlife Service, being pro-moted to aquatic biologist in 1945 and the following year to information specialist. Her salary was growing, her responsibilities expanding, and she was given more opportunities to travel and do field work. She summarized her job duties in the following response on a federal questionnaire con-cerning women in government: "My job consists of general direction of the publishing program of the Service—working with authors in planning and writing their manuscripts, reviewing manuscripts submitted, and overseeing the actual editing and preparation of the manuscript for the printer. I have a staff of six assistants.... It is really just the work of a small publishing house" (quoted in Brooks 1972: 99).

Despite her success at the FWS, Rachel was still frustrated that she could not pursue what she sensed as her true calling. In 1947, she wrote to friend and amateur ornithologist Ada Govan:

> No, my life isn't at all well ordered and I don't know where I'm going! I know that if I could choose what seems to me the ideal existence, it would be just to live by writing. But I have done far too little to dare risk it. And all the while my job with the Service grows and demands more and more of me, leaving less time that I could put on my own writing. And as my salary in-creases little by little, it becomes even more impossible to give it up! (quoted in Lear 1997: 130)

Govan, with whom Rachel had become acquainted while she was re-searching an article about bird banding, was the author of *Wings at My Window*, the story of the Woodland Bird Sanctuary. The two women dis-covered that, in addition to a shared love of nature, they were both writ-ers struggling to make a living from their writing.

Striving to make the best of her job situation, Rachel had an idea for a project that would give added meaning to her work and give her further opportunity to write and travel. She proposed a series of booklets focusing on individual national wildlife refuges as well as ecology in general. The refuges had been established beginning in the early 1930s as a result of the Migratory Bird Conservation Act, and by the mid-1940s, there were about three hundred refuges nationwide. In the postwar period, a population explosion was occurring in the United States, and with it, a building boom that was threatening the wild areas of the country. Highways linked the endless sprawl of suburbs as veterans of the war had returned home to start new families. Carson's booklets, which belonged to the series she titled *Conservation in Action*, were, therefore, timely publications that would encourage readers to support the conservation of wild areas.

The refuge guides are concerned mostly with birds, including information on migration patterns, feeding habits, and reproduction of the bird species that a visitor might encounter at each refuge. Carson was put in charge of the series, writing four of the twelve booklets herself and coauthoring a fifth. In a mission statement that prefaces each booklet, she explained, "Wild creatures, like men, must have a place to live. As civilization creates cities, builds highways, and drains marshes, it takes away, little by little, the land that is suitable for wildlife. And as their space for living dwindles, the wildlife populations themselves decline" (quoted in Sterling 1970: 109). In expressing the philosophy behind the series, Carson recognized the interconnection of all creatures, including humans, with their environment, writing, "the preservation of wildlife and wildlife habitat means also the preservation of the basic resources of the earth, which men, as well as animals, must have in order to live. Wildlife, water, forests, grasslands—all are parts of man's essential environment; the conservation and effective use of one is impossible except as the others also are conserved" (quoted in Brooks 1972: 101).

To complete the brochures, Rachel visited a number of refuges for field research. In April 1946, she and Shirley Briggs went to the waterfowl refuge on Assateague Island in Chincoteague, Virginia, that had recently been added to the system. This refuge, along with a number of others along the eastern coast of the United States, offered migrating birds a protected place to stop on their long journeys. The ultimate survival of these birds, some migrating from the southern tip of South America all the way to northern Greenland, depended on these areas along what became known as the Atlantic flyway. In the booklet *Chincoteague*, Carson wrote, "Once there were plenty of natural hostelries for the migrants. That was before our expanding civilization had drained the marshes, polluted the waters, substituted resort towns for wilderness. That was in the days when hunters were few" (quoted in McCay 1993: 35).

In September 1946, Rachel visited the Parker River Refuge, which she considered the most important refuge in New England. Located along the northern coast of Massachusetts, it was, like Chincoteague, part of the Atlantic flyway and provided a haven for some of the same species. Many residents did not like having the refuge in their county because it threatened the soft-shelled clam industry. Rachel and Kaye Howe, who was travelling with her, encountered some animosity from the local people because of their efforts to publicize the refuge, and Kay believed that her film was destroyed by a resident who had offered to get it developed for her.

In February 1947, Carson travelled to three refuges in North Carolina along the Atlantic flyway that protected the whistling swan. The booklet resulting from her visit to one of these refuges, Mattamuskeet, "reflects the confident writing of a mature scientist who knows her subject, her audience, and her public mission to inform. It also exhibits Carson's understanding of the intricate ecology of a wildlife habitat and her desire to communicate the importance of these ecological relationships" (Lear 1998: 1). In addition to providing accurate factual information regarding the marshland environment, the habits of the wildlife, and the activities of the refuge management, she included literary descriptions that seem to transcend science. For example, in describing the sound of the whistling swans, she wrote, "Underlying all the other sounds of the refuge is their wild music, rising at times to a great, tumultuous crescendo, and dying away again to a throbbing undercurrent" (quoted in Lear 1998: 44).

The booklets Chincoteague, Parker River, and Mattamuskeet were all published in 1947. In the fall of that year, Rachel took a train west with Kay Howe to spend a month gathering material for Guarding Our Wildlife Resources, a booklet in the Conservation in Action series that would provide an overview of the Fish and Wildlife Service's efforts to protect migratory birds, endangered animals, fish, and other wildlife. They visited the Red Rock Lakes Refuge that protected trumpeter swans on the border of Montana and Idaho near Yellowstone Park. They also went to the National Bison Refuge in Montana, the salmon hatcheries on the Columbia River in Oregon, and the Bear River Refuge near Salt Lake City on the border of the Central and Pacific flyways. During their trip, they had the opportunity to drive to Agate Beach in Oregon, the first time either one had seen the Pacific Ocean. Guarding Our Wildlife Resources was published in 1948, and Bear River, cowritten with Vanez Wilson, appeared in 1950.

The Conservation in Action booklets that Carson wrote are considered fine examples of nature writing even though the information in them may be outdated. She thoroughly researched her subjects and then wrote in her characteristic literary style that was unusual in government publications. Her introduction to ecological concepts, as she addressed issues such as pollution and the relationship of wildlife to the environment, was quite remarkable for that era. At a time when ecology was not widely appreciated, her insights were revolutionary. Although Rachel considered the *Conservation in Action* project to be her most fulfilling work with the FWS, she still had the nagging feeling that she had not yet fulfilled her life's true purpose.

# Chapter 5

# WRITING *THE SEA AROUND US*

## SECOND THOUGHTS ABOUT ANOTHER BOOK

Rachel Carson had begun to reconsider writing another book while working on *Conservation in Action*. While she was visiting the Parker River Refuge in 1946, she met with Dr. Henry Bigelow, the former director of the Woods Hole Oceanographic Laboratory and oceanographic curator at Harvard University's Museum of Comparative Zoology. He encouraged her to write a book about the history of the sea based on new oceanographic research conducted during the war. She started gathering material and drawing up an outline for a book that she tentatively titled *Return to the Sea*.

As with her first book, Rachel worked on *Return to the Sea* at night and on weekends, never feeling that she had the time for the project. Although her full-time job seemed to interfere with her writing, her position did provide her with access to the information she needed, allowing her to utilize previously classified government material about oceanography. In 1948, she wrote to William Beebe, "The book I am writing is something I have had in mind for a good while. I have had to wait to undertake it until at least a part of the wartime oceanographic studies should be published" (quoted in Brooks 1972: 110). During World War II, vast amounts of information were collected concerning the ocean and the environment in general. This research would have taken decades longer had it not been for the urgency created by the war. Successful naval operations required a thorough knowledge of the ocean and the effect of tides, waves, and currents on ships and submarines. The rapid development of scientific instruments during World War II revolutionized the science of oceanography. Wave recorders analyzed the origin and speed of waves on the surface, while the echo sounder revealed what lay at the bottom of the sea. Underwater instruments called hydrophones, used for detecting sounds, were set up along the coast of the United States to listen for enemy submarines. They also revealed the surprisingly noisy world under the surface that was populated by a broad range of sea creatures.

*Return to the Sea* would be written, like *Under the Sea-Wind*, for the general reader. Rachel wanted to share her fascination with and love of the sea so that her readers would appreciate the marine environment and to write a book that she herself would have enjoyed reading earlier in her life. But she did not want it to be just "another 'introduction to oceanography' " (quoted in Brooks 1972: 124). She wrote to William Beebe, "I am much impressed by man's dependence upon the ocean, directly, and in thousands of ways unsuspected by most people. These relationships, and my belief that we will become even more dependent upon the ocean as we destroy the land, are really the theme of the book and have suggested its tentative title, 'Return to the Sea' " (110). In her application for the Eugene F. Saxton Memorial Fellowship, which provided support for promising writers, Rachel summarized *Return to the Sea* as "an imaginative searching out of what is humanly interesting and significant in the life history of the earth's ocean; and the answering of questions thus raised in the light of the best available scientific knowledge" (quoted in Lear 1997: 162).

Because Carson had been dissatisfied with Simon & Schuster's handling of *Under the Sea-Wind* and partly blamed the publisher for its poor sales, she decided to find a new publisher for her next book and hired a literary agent. In May 1948, she chose Marie Rodell, who was starting her own literary agency after a successful career as an editor and mystery writer. Rachel soon became good friends with the outgoing and vivacious Rodell despite their personality differences and found the energetic Rodell to be a constant source of encouragement.

While Rachel was working on the *Conservation in Action* series, she had little time for freelance writing. She did manage to write one article during 1947, about an invasion of microscopic organisms called *Gymnodinium* that was killing a large number of fish in the Gulf of Florida. By the time "The Great Red Tide Mystery" appeared in the February 1948 issue of *Field and Stream*, Rachel defined herself more by her moonlighting as a writer than by her full-time job. In a letter to ornithologist Robert Cushman Murphy, curator of the bird department at the American Museum of Natural History, she introduced herself as "a marine biologist whose actual profession is writing rather than biology" (quoted in Lear 1997: 154).

Over the summer of 1948, Rachel compiled a complete outline of *Return to the Sea* that Rodell could submit to publishers and then worked on a sample chapter about the formation of islands. The first publisher that received the material, William Sloan Associates, rejected the book proposal, claiming they could not make a decision based on an outline and a sample chapter. Some unrelated bad news was to follow this rejection. Mary Scott Skinker was dying of cancer. Rachel traveled to Chicago to see the woman who had had such a profound influence on her life and was devastated by her death on December 19, 1948, at the age of fifty-seven. Marie Rodell was one of the few people whom Carson told about the event.

In April 1948, Philip Vaudrin, the editor of Oxford University Press, contacted Marie to express interest in Carson's book proposal, and on June 28, 1949, Rachel signed a contract, received an advance of $1,000, and was given an earlier deadline than anticipated—March 1, 1950. At about the same time, her job responsibilities increased. During the summer of 1949, Bert Walford was promoted to head of the Fish and Wildlife Service, and Rachel was promoted to his vacated position of biologist and chief editor. She received a raise, an increase in grade, and all of Walford's duties, but she was denied his administrative position or grade.

## RESEARCHING *RETURN TO THE SEA*

That summer Rachel visited an area in Florida that would become Everglades National Park to do some research for an FWS publication and a bit of her own research as well. She was impressed with the similarities between the Everglades and the ocean. In *The Sea Around Us,* which was the final title of her second book, she wrote:

> Far in the interior of the Florida Everglades I have wondered at the feeling of the sea that came to me—wondered until I realized that here were the same flatness, the same immense spaces, the same dominance of the sky and its moving, changing clouds; wondered until I remembered that the hard rocky floor on which I stood, its flatness interrupted by upthrust masses of jagged coral rock, had been only recently constructed by the busy architects of the coral reefs under a warm sea. (Carson 1951: 101)

Rachel and Shirley Briggs went deep into the Everglades with a local guide who had never taken women aboard his "glades buggy," a tractor-like vehicle with six pairs of large wheels and an exposed engine that blew heat on them throughout their exploration of the swampy area. They also endured the discomfort of torrential rain and swarms of mosquitoes.

While in Florida, Rachel went on a diving trip with Dr. F. G. Walton-Smith, a biologist at the Miami Marine Laboratory. William Beebe had suggested that Rachel go diving to gather firsthand information on the topic she was writing about in her book. Due to uncooperative weather, however, her diving experience was not outwardly very successful, and she only had a brief exposure to the undersea world of the reefs off the Florida Keys. At fifteen feet below the surface, which was as far as she descended, it was difficult to see much in the murky waters or move about freely against the strong currents. She was profoundly moved, however, by the perspective she gained from being underwater and looking up to the surface. She later wrote to Beebe, "But the difference between having dived—even under those conditions—and never having dived is so tremendous that it formed one of those milestones of life, after which everything seems a little different" (quoted in Lear 1997: 169). While she was still in Florida, her mother called to tell her that she had received the Saxton Fellowship of $2,250, which would allow her to take time off from work to write her book.

As part of her research, Rachel wanted to go on a trip aboard the FWS vessel, the *Albatross III*. Officials were concerned, however, about the propriety of a single woman being alone among a crew of more than fifty men. No other woman had ever been on the *Albatross III*. As a solution, Rachel asked Marie Rodell to accompany her. Marie joked that she would write an article about her experience titled "I Was a Chaperone on a Fishing Boat." As soon as Rachel and Marie boarded the ship in Woods Hole in July 1949, the officers, who felt a bit uneasy about having women aboard, cautioned them about the dangers of the ship, which included being washed overboard or injured by fishing gear, and warned them about the unpleasantness of seasickness and bad food. The two women, whose friendship was strengthened by the voyage, endured all the inconveniences in a positive manner. They even managed to get a good night's sleep after a couple of days, once they became accustomed to the noise of the trawl as it was dropped into the water, dragged over the ocean bottom, and hauled back on deck at all hours of the day and night.

The ten-day voyage aboard the *Albatross III* to Georges Bank, an area of the Atlantic Ocean known for fishing that lies two hundred miles east of Boston and south of Nova Scotia, was organized to study the growing scarcity of popular commercial fish such as cod. Rachel's mission, in addition to research for her book, was to gather information for future FWS publications about the conservation efforts of the *Albatross III* as it conducted a census of the fish population. One of Rachel's favorite parts of the trip was using the echo sounder to analyze the immense canyons that characterized the undersea world of Georges Bank. She studied specimens under a microscope in the ship's laboratory with the other biologists and observed the crew's use of depth recorders and sonar in the operations room.

Rachel was impressed by the creatures that were hauled aboard the deck of the ship by the trawl. In addition to the usual fish, crabs, and sponges, there were often more exotic specimens that she had never seen before. Although most of the fish that were caught were thrown back in the ocean after they were counted and measured, Rachel was greatly disturbed when the crew shot some of the sharks for sport. She gained a new perspective from being aboard a ship and later explained, "There is something deeply impressive about the night sea as one experiences it from a small vessel far from land. When I stood on the afterdeck on those dark nights, on a tiny man-made island of wood and steel, dimly seeing the great shapes of waves that rolled about us, I think I was conscious as never before that ours is a water world, dominated by the immensity of the sea" (quoted in Lear 1998: 154).

The Saxton Fellowship allowed Rachel to take a leave of absence from her job. She took off the entire month of October 1949, several weeks in December, and two months in 1950 to focus on completing the book, which was due by March. She often found herself going back to the office for emergencies and bringing work home with her, however, hampering her efforts. She was also distracted by an idea for a completely unrelated project. Having admired the bird paintings of artist Louis Agassiz Fuertes, she envisioned a book that would be part art catalog and part biography. She offered this idea to Philip Vaudrin at Oxford, who at first was interested but then declined because he thought the costs of reproducing the paintings would be too high.

Rodell then contacted Paul Brooks, editor-in-chief of general books at Houghton Mifflin in Boston, about Rachel's new book idea. Paul Brooks was an amateur naturalist and ornithologist and had edited the field guides of his friend Roger Tory Peterson. He told Carson that he would consider her proposal, but he also had in mind another project that she might want to undertake—a guidebook to creatures of the shore for the general public. Another editor at Houghton Mifflin, Rosalind Wilson, had suggested the need for such a book after returning from Cape Cod one weekend. While she had been walking on the beach with a group of friends, they came upon some horseshoe crabs that they thought were stranded and, believing they were helping the creatures, put them back in the ocean. They did not realize that the crabs intentionally went ashore to lay their eggs. Wilson thought a book should be written for the layperson that "would dispel such ignorance" (Brooks 1972: 152). The seashore guide and the Fuertes book would both have to wait, however, because Rachel had to finish *Return to the Sea* by March 1, 1950.

Rachel never really liked the title *Return to the Sea* and struggled with Marie in early 1950 to come up with a new title. They considered *The Story of the Ocean, Story of the Sea, Empire of the Sea,* and *Sea without End* but ultimately rejected all of these, settling on *The Sea Around Us,* which Rachel had thought of earlier. She was unable to meet the March deadline, so Marie negotiated an extension, and the manuscript was finally submitted at the end of June 1950.

## AN OVERVIEW OF *THE SEA AROUND US*

Rachel Carson intended for *The Sea Around Us* to give readers the perspective that she had gained on the *Albatross III*—that the earth is truly "a water world, a planet dominated by its covering mantle of ocean, in which the continents are but transient intrusions of land above the surface of the all-encircling sea" (Carson 1951: 15). She blended the latest oceanographic research with the mythology of the sea and shared personal anecdotes such as her voyage aboard the *Albatross.* Throughout the book, historical background is provided on the exploration of the sea by explorers such as Magellan, scientists such as her friend William Beebe, and the members of the U.S. Navy whose wartime research provided Carson with much of the material contained in her book.

Carson consulted more than a thousand printed sources in the writing of *The Sea Around Us.* She also corresponded with numerous oceanographers and other scientists all over the world, including Henry Bigelow who had initially encouraged her to write the book, renowned ornithologist Robert Cushman Murphy, and Thor Heyerdahl, author of *Kon-Tiki,* an autobiographical account of a journey across the Pacific on a raft. Although she relied heavily on scientific fact, she acknowledged that much about which she wrote, including the origin of the moon, the earth, and life itself, was based on conjecture.

The book is divided into three parts with Part I, "Mother Sea," being the longest section of the book. After an introductory chapter focusing on the origin of the earth and sea, several chapters analyze the nature of the sea from the surface waters to the ocean floor. The intricate food chain that exists in the ocean is explained, as well as the seasonal changes that occur in the surface waters, and the strange creatures that inhabit the ocean depths are vividly depicted. This first section also includes "The Birth of an Island," which would soon appear on its own in the *Yale Review*. This chapter explains how islands develop and sometimes die and describes the delicate ecosystems created by isolated islands. The section ends with speculation on the ever-changing relationship between land and ocean brought about by rising seas.

Part II, "The Restless Sea," is about forces that affect the ocean, including wind, the sun and moon, and the earth's rotation, and about the power of the sea as embodied in waves, tides, and currents. "Wind and Water" explains the nature of waves, how winds create waves, how waves can cause destruction but, at the same time, create the beautiful shoreline. Also included in this chapter is a report on how the measurement and study of waves was important to military operations during World War II. The permanent currents of the ocean are described in the next chapter, as well as cosmic forces such as the sun, moon, and wind that create currents such as the Gulf Stream. Finally, the mysterious and powerful force of gravitational pull, which creates the tides, affecting every drop of water in the ocean, is discussed in "The Moving Tides."

The final part of the book, "Man and the Sea about Him," describes how the ocean has influenced and continues to influence human beings. Changes in ocean currents such as the Gulf Stream, for example, produce seasonal variations in climate as well as long-range changes, such as ice ages, that have altered the course of history. The ocean's mineral resources, especially petroleum, salt, and gold, have proven valuable to humankind throughout the ages. The development of navigational methods since ancient times has allowed people to explore the sea, and, as a result, "through many voyages undertaken over many centuries, the fog and the frightening obscurity of the unknown were lifted from all the surface of the Sea of Darkness" (Carson 1951: 211).

As in *Under the Sea-Wind*, the major themes of *The Sea Around Us* are the interdependence of all living things and the relationship of creatures to their environment. These themes are evident when the microscopic organisms of the surface waters are linked to other creatures much further down. In Carson's words, "What happens to a diatom in the upper, sunlit strata of the sea may well determine what happens to a cod lying on a ledge of some rocky canyon a hundred fathoms below, or to a bed of multicolored, gorgeously plumed seaworms carpeting an underlying shoal, or to a prawn creeping over the soft oozes of the sea floor in the blackness of mile-deep water" (Carson 1951: 19).

The theme of interdependence is especially apparent in "The Birth of an Island." In this chapter, islands are characterized as small self-contained ecosystems in which humans have had a negative influence. Carson asserted that man "has written one of his blackest records as a destroyer on the oceanic islands. He has seldom set foot on an island that he has not brought about disastrous changes" (Carson 1951: 93). Later, she lamented, "The tragedy of the oceanic islands lies in the uniqueness, the irreplaceability of the species they have developed by the slow processes of the ages. In a reasonable world men would have treated these islands as precious possessions, as natural museums filled with beautiful and curious works of creation, valuable beyond price because nowhere in the world are they duplicated" (96–97).

Although *The Sea Around Us* shares themes with Carson's first book, the point of view differs. In *Under the Sea-Wind*, the stories were told through the eyes of the sea creatures themselves, whereas in *The Sea Around Us*, Carson explained some of the mysteries of the ocean in first person prose. She frequently drew readers in by directly addressing them as "you" or "we." For example, she suggested that readers contemplate the origin of the moon the "next time you stand on a beach at night" (Carson 1951: 5). Later, she intrigued readers as she linked the present to the past, speculating, "The surf that we find exhilarating at Virginia Beach or at La Jolla today may have lapped at the base of Antarctic icebergs or sparkled in the Mediterranean sun, years ago, before it moved through dark and unseen waterways to the place we find it now" (150).

Carson also appealed to readers by implying that human beings are in some distant way related to the creatures of the sea because the sea was long ago the home of the earliest ancestors of man. She wrote:

> Fish, amphibian, and reptile, warm-blooded bird and mammal—each of us carries in our veins a salty stream in which the elements sodium, potassium, and calcium are combined in almost the same proportions as in sea water. This is our inheritance from the day, untold millions of years ago, when a remote ancestor, having progressed from the one-celled to the many-celled stage, first developed a circulatory system in which the fluid was merely the water of the sea. (Carson 1951: 13–14)

She also reinforced the connection of past to present by reminding readers that many areas of land were once covered by the sea and may be again someday. Her book concludes with these words: "For the sea lies all about us.... In its mysterious past it encompasses all the dim origins of life and receives in the end, after, it may be, many transmutations, the dead husks of that same life. For all at last return to the sea—to Oceanus, the ocean river, like the ever-flowing stream of time, the beginning and the end" (216).

In the process of writing *The Sea Around Us*, Carson experienced what she later described as "the thing that happens when one has finally established such unity with one's subject matter that the subject itself takes over and the writer becomes merely the instrument through which the real act of creation is accomplished" (quoted in Freeman 1995: 148). She began gathering information for her book in 1946, although she later said, "People often ask me how long I worked on *The Sea Around Us*. I usually reply that in a sense I have been working on it all my life, although the actual writing of the book occupied only about three years" (quoted in Brooks 1972: 110). *The Sea Around Us* was an outgrowth of Rachel's love of the ocean that had begun in childhood, long before she had even laid eyes on it.

# Chapter 6

# THE WAVE OF REACTION
# TO *THE SEA AROUND US*

## AWAITING PUBLICATION DAY

Although Rachel Carson had completed her manuscript for *The Sea Around Us* and returned to work full-time, she needed some extra income until her book was published and began generating royalties. Marie Rodell tried to sell individual chapters to magazines but was rejected by almost twenty publications, including the *Atlantic Monthly*, which had published "Undersea" in 1937. This initial negative reaction ended when William Shawn, an editor at the *New Yorker*, expressed interest in some of the chapters from *The Sea Around Us* because he felt that it was destined to become a great book. Just days later, the *Yale Review* bought "The Birth of an Island" and published it in the September 1950 issue.

Unfortunately for Rachel, these successes were overshadowed by the diagnosis of a tumor in the same breast from which a smaller cyst had been removed four years earlier. She outwardly expressed a casual attitude of indifference toward this condition, but in a letter to Edwin Way Teale, her friend and fellow nature writer, she revealed that she felt "as writers should, a sense of urgency and passing time—and so much to say!" (quoted in Lear 1997: 185). She had an operation to remove the tumor on September 21, 1950, and, because it was not identified as malignant, no further treatment was recommended.

In an effort to recuperate fully from the operation and also alleviate the exhaustion she had been suffering, she spent a week in Nags Head, North Carolina. It was a reflective time as indicated by her notebook in which she wrote, "Time itself is like the sea, containing all that came before us, sooner or later sweeping us away on its flood and washing over and obliterating the traces of our presence" (quoted in Lear 1998: 126). She spent time thinking about the seashore guide that editor Paul Brooks had proposed to her. Instead of a guide to certain seashore sites as originally conceived, she envisioned a book that would focus on the ecology of shore life. While in North Carolina, she also completed an application for a Guggenheim Fellowship, which was eventually awarded to her in March 1951, allowing her to take a year's leave of absence from her job to write. She explained in this application that she felt her special talent as a writer was "the interpretation of scientific findings in terms that give them reality and meaning for the non-scientific reader" (quoted in Lear 1997: 187).

When Rachel returned to work in mid-October, she was concerned about a delay in the publication of *The Sea Around Us*. The *New Yorker*, which was interested in publishing some of her chapters, also wanted to have the publication of the book postponed one year, but Carson refused. She was afraid that the outbreak of the Korean War would have a negative impact on the sales of her book, just as World War II had done for *Under the Sea-Wind*. To add to her anxiety, she was worried that the Fish and Wildlife Service might relocate her again.

Marie was becoming more successful in her efforts to sell some of the chapters to magazines. In addition to "The Birth of an Island," which appeared in the *Yale Review*, *Science Digest* eventually printed the chapter "Wealth from the Salt Seas," and *Vogue*, departing from its usual range of topics, bought "The Global Thermostat." "The Birth of an Island" won the Westinghouse Science Writing Prize given by the American Association for the Advancement of Science for the best science article published nationally in 1950. Although Rachel appreciated the thousand dollars that came with the award, she was more excited about the official recognition given to her work.

Although the *New Yorker* did not generally publish scientific material, editor William Shawn eventually bought nine of the fourteen chapters in *The Sea Around Us*, condensed them, and published them as a three-part "Profile of the Sea" in June 1951, the month before the publication of her book. The "Profile" column usually highlighted a famous person and had never featured a place or thing. Rachel received $7,200 which was more than her annual salary. More significantly, this preview of her book helped it to become a best-seller. After the "Profile of the Sea" appeared in the *New Yorker*, the magazine was inundated with letters of praise for Carson's work.

## THE INITIAL REACTION

On July 1, 1951, the day before *The Sea Around Us* was published, it received a positive review on the front page of the *New York Times Book Review* when Jonathan Norton Leonard complimented the book for being written with the "precision" of a scientist and the "style and imagination" of a poet (Leonard 1951: 1). Other critics praised Carson as "both scientist and stylist" and "one of those very rare scientists who can also write magnificently" (*Kirkus* 1951: 285; Jackson 1951: 14). Harry Ellis commented that *The Sea Around Us* "offends neither the natural scientist nor the poet," while Francesca La Monte called it "one of the most beautiful books of our time" (Ellis 1951: 7; La Monte 1951: 3). Months later, after the rerelease of *Under the Sea-Wind*, a review in the *New York Times* declared, "Once or twice in a generation does the world get a physical scientist with literary genius. Miss Carson has written a classic in *The Sea Around Us*" (*New York Times* 1952: IV, 8).

*The Sea Around Us* appeared on the *New York Times* best-seller list by the end of July—remarkable for a book on a topic such as this from a relatively unknown writer. It went to the top of the list on September 9, 1951, and was a bestseller for eighty-six weeks, spending thirty-two weeks in first place. By early November, *The Sea Around Us* had sold one hundred thousand copies. During the Christmas shopping season that year, it was selling about four thousand copies per day, and the publisher had a difficult time printing enough copies to keep bookstore shelves stocked. It was abridged by *Reader's Digest*, was offered as a Book-of-the-Month Club alternate selection, and was published in thirty-two languages. Before the end of 1952, it had sold a quarter of a million copies.

## WHY *THE SEA AROUND US* BECAME A BESTSELLER

Even Rachel herself did not understand why *The Sea Around Us* sold so well, telling Marie Rodell, "I simply can't understand the way the public has gotten so mad about the sea!" (quoted in Lear 1997: 208). The condensed version that appeared in the *New Yorker* certainly helped, but this alone cannot explain the book's popularity. Although in the society of the early 1950s, environmental topics were not of much interest, books related to the sea sold well. The fiction and nonfiction best-seller lists in 1951 included two novels with naval themes, *From Here to Eternity* by James Jones, set in Hawaii, and *Caine Mutiny* by Herman Wouk; Thor Heyerdahl's *Kon-Tiki* was number one on the nonfiction list at the time.

With the development of atomic weapons by both the United States and the Soviet Union that had followed the end of World War II, fear of nuclear devastation gripped the nation. Although citizens were preoccupied with the Cold War between the former allies that had become adversaries as well as the conflict in Korea, many still found time to read *The Sea Around Us*. Amid a tense political atmosphere, Americans began to appreciate the ocean more than ever as a neutral territory that isolated their country. Many people found a comforting escape in reading Carson's book, which covered a topic spanning millions of years making current events seem less significant. As Linda Lear explained, "A nation fearful of the escalating nuclear arms race, made nervous by Joseph McCarthy's hunt for domestic Communists, and reluctant to send their sons to fight a war in a far-off Pacific nation like Korea found in *The Sea Around Us* a longer perspective on their problems and a larger dimension by which to measure human achievement" (Lear 1997: 205). Lear also stated, "By focusing on the immutable forces of nature, *The Sea Around Us* calmed atomic fears" (Lear 1997: 220).

Although Carson's first book, *Under the Sea-Wind*, was similar to *The Sea Around Us* in its timeless perspective, the mood of the country was more apprehensive in the early 1950s. During the Cold War, there was neither the huge deployment of troops that characterized World War II nor the widespread activity on the home front to contribute to the war effort. In short, people were waiting in suspense and needed something to take their minds off their worries. Because of this situation, Carson's second book was a huge success whereas her first had been a disappointment.

Carson gave us a radical perspective on time in *The Sea Around Us* as she commented that man "often forgets the true nature of his planet and the long vistas of its history, in which the existence of the race of men has occupied a mere moment of time" (Carson 1951: 15). When she later explained the force of tidal friction that is slowly pushing the moon away from the earth, diminishing the moon's influence on the tides, she wrote, "All this, of course, will require time on a scale the mind finds it difficult to conceive, and before it happens it is quite probable that the human race will have vanished from the earth" (158).

In a speech given at the *New York Herald Tribune* Book and Author Luncheon held at the Astor Hotel in New York City on October 16, 1951, Carson, the featured speaker, said, "The sea is a place where one gets a sense of the great antiquity of the earth. It seems changeless; but it is always changing. It links the dim beginnings of time with the present" (quoted in Lear 1998: 78). At a benefit luncheon for the National Symphony Orchestra, she said, "When we contemplate the immense age of earth and sea, when we get in the frame of mind where we can speak eas-ily of 'millions' or 'billions' of years, and when we remember the short time that human life has existed on earth, we begin to see that some of the worries and tribulations that concern us are very minor" (88–89).

Commenting on her fan mail, Carson said, "It has come to me very clearly through these wonderful letters that people everywhere are des-perately eager for whatever will lift them out of themselves and allow them to believe in the future" (quoted in Lear 1998: 89). Years later, she wrote a letter to her friend Dorothy Freeman about her troubles sleeping following the assassination of President John Kennedy and how she found solace in reading her first book *Under the Sea-Wind* again, just as others had been comforted by her second book amid Cold War fears. She re-counted:

> Finally I picked up *Under the Sea-Wind!* And somehow it was right. A chapter or two a night relaxed me and let me sleep.... Of course it is the elemental nature of the subject matter, its timelessness, beside which human problems and even human tragedy fall into perspective. But out of this experience I un-derstood for the first time what various people have told me about reading that, or *The Sea*...in time of trouble. (quoted in Freeman 1995: 499)

After the publication of *The Sea Around Us*, Rachel was invited to write the jacket notes for a recording of Claude Debussy's *La Mer* (which means "the sea") conducted by Arturo Toscanini. She had always enjoyed music, a love she had inherited from her mother, but she hardly felt like a competent music critic. Her comments go beyond the music itself and re-flect her appreciation for the timelessness and power of the sea, which can alter the perspective of mankind:

> What is this sea, and wherein lies its power so greatly to stir the minds of men? What is the mystery of it, intangible, yet insep-arably its own? Perhaps part of the mystery resides in its hoary antiquity, for the sea is almost as old as earthly time.... Or per-haps the spirit of the sea resides in the implacable, inexorable power by which it draws all things to it, by which it over-whelms and devours and destroys.... Or perhaps the mystery is the mystery of life itself—of life that began as a primordial bit of protoplasm adrift in the surface waters of the ancient seas. (quoted in Lear 1998: 87)

Although Carson herself was astonished by the popularity of her book, she said, "Many people have commented with surprise on the fact that a work of science should have a large popular sale. But this notion that 'science' is something that belongs in a separate compartment of its own, apart from everyday life, is one that I should like to challenge" (quoted in Lear 1998: 91).

In a special edition of *The Sea Around Us* published in 1989, Ann Zwinger asserted in her introduction that "it was *The Sea Around Us* that started people thinking, that laid the groundwork for stronger medicine to come" (Zwinger 1989: xxv). The "stronger medicine" to which she referred is Carson's most famous work, *Silent Spring*, considered by some to have started the environmental movement. Although *The Sea Around Us* is not as well-known as *Silent Spring*, it was also significant in raising the environmental consciousness of society. Zwinger continued:

> At first glance, *The Sea Around Us* does not seem the kind of book of which revolutions are made.... It rattles no swords, is not strident or confrontational. Its potency lies in the charm and skill of the writing, its erudition and rich organization of facts, and its personal reticence—how quietly it captivates our attention. Before we know it we are charmed into learning about the wonders of the ocean, then into a deep awareness of not only their health but how it affects that of the whole natural world. Through sharing Carson's research, we become acutely sensitive to the interdependence of life. (xxv–xxvi)

## AWARDS FOR *THE SEA AROUND US*

Aside from critical acclaim and impressive sales, Rachel Carson and her book were showered with accolades. While *The Sea Around Us* was named an "outstanding book of the year" by the *New York Times*, the Associated Press named its author "woman of the year in literature." The Garden Club of America awarded Carson the Frances K. Hutchinson Medal recognizing her achievement in the area of conservation. The Philadelphia Geographical Society awarded her its Henry G. Bryant Medal, and the New York Zoological Society gave her the Gold Medal for her work. The Limited Editions Club nominated her as one of ten living American writers whose books published between 1929 and 1954 seemed "most likely to survive as classics," and in May 1954, she was awarded the Club's Silver Jubilee Medal in New York City.

*The Sea Around Us* won the 1951 National Book Award for nonfiction. At the award ceremony, the following citation was read: "Rachel L. Carson's *The Sea Around Us* brings to the attention of the public a hitherto unconsidered field of scientific inquiry of great importance to the spiritual and material economy of mankind. It is a work of scientific accuracy presented with poetic imagination and such clarity of style and originality of approach as to win and hold every reader's attention" (quoted in Lear 1997: 218). The prestigious John Burroughs Medal, named after a popular American nature writer of the late nineteenth and early twentieth centuries, and given each year for an outstanding book on the subject of natural history, was awarded to *The Sea Around Us* in 1952. For Rachel, this was the most coveted prize she could receive as an acknowledgment of the quality of her work. In accepting this award, she emphasized the importance of nature writing in a modern society where humans have become increasingly alienated from nature, asserting:

> I myself am convinced that there has never been a greater need than there is today for the reporter and interpreter of the natural world. Mankind has gone very far into an artificial world of his own creation. He has sought to insulate himself, in his cities of steel and concrete, from the realities of earth and water and the growing seed. Intoxicated with a sense of his own power, he seems to be going farther and farther into more experiments for the destruction of himself and his world. (quoted in Lear 1998: 94)

Her speech also encouraged nature writers to realize that general readers, not just fellow scientists, thirsted for knowledge on the subject of nature.

Carson was elected a Fellow of the Royal Society of Literature in England and to the National Institute of Arts and Letters in the United States, only the second woman to receive this honor for literary merit. She was given an honorary membership in Theta Sigma Phi, the national fraternity of women in journalism, and was awarded honorary doctorates by her alma mater, Pennsylvania College for Women, as well as Oberlin College, Drexel Institute of Technology, and Smith College. Although asked by many other educational institutions to receive honorary degrees, she only accepted these four.

*Under the Sea-Wind* was rereleased by Oxford on April 13, 1952, and also became a bestseller, making its first appearance on the *New York Times* best-seller list at tenth place while *The Sea Around Us* held second place. This book was chosen as an alternate selection by the Book-of-the-Month Club that June. Henry Beston, a writer that Rachel greatly admired, gave a glowing review of *Under the Sea-Wind* in the *Freeman*, writing, "It is Miss Carson's particular gift to be able to blend scientific knowledge with the spirit of poetic awareness, thus restoring to us a true sense of the world" (Beston 1952: 100). On April 27, the *New York Times* called the popularity of these two books a "publishing phenomenon rare as a total solar eclipse" (*New York Times* 1952: IV, 8). As royalties from *The Sea Around Us* and her rereleased first book accumulated, Rachel returned the Guggenheim Fellowship check sent to her in March, feeling that other worthy applicants needed the money more than she did, and asked that no further stipend be sent to her. She resigned altogether from the Fish and Wildlife Service in June 1952.

A film version of *The Sea Around Us* produced by Irwin Allen at RKO won an Academy Award for best full-length documentary of 1953, but Carson disliked the project because of the many scientific inaccuracies the film contained. Considering the film to be amateurish, sensationalized, and melodramatic, she was particularly bothered by the way in which the sea creatures were anthropomorphized, a subject that she had had such strong feelings about in her first book. The only thing that she thought her book had in common with the film was the title. Because the movie studio had ignored Carson's advice on how to make the film, she became skeptical about any future projects based on her work unless she had complete editorial control.

## MISPERCEPTIONS ABOUT CARSON

Although she became a fairly well-known celebrity, at first many people held misperceptions about Rachel Carson. One letter to the *New Yorker*, written by someone who thought that Rachel Carson was a pseudonym, stated, "I assume from the author's knowledge that he must be a man" (quoted in Lear 1997: 206). Later, after *The Sea Around Us* was published without a photo of the author on the book jacket, some readers also thought she was a man. Others thought she must be very old because of the time it must have taken to collect all the information in the book. Shirley Briggs drew a cartoon titled "Rachel as her readers seem to imagine her" that depicted her friend as an Amazon warrior. Many people were surprised when they finally saw the petite, slender, and relatively young Carson, whose picture appeared on the cover of the *Saturday Review of Literature* in July 1951.

Almost all the male book reviewers questioned what Carson might look like in their reviews. For example, the *New York Times* writer Jonathan Norton Leonard closed his review with this statement: "It's a pity that the book's publishers did not print on its jacket a photograph of Miss Carson. It would be pleasant to know what a woman looks like who can write about an exacting science with such beauty and precision" (Leonard 1951: 1). A *Boston Globe* interview, which mistakenly referred to her first book as *Under the Seaweed,* also focused on her appearance: "Would you imagine a woman who has written about the seven seas and their wonders to be a hearty physical type? Not Miss Carson. She is small and slender, with chestnut hair and eyes whose color has something of both the green and blue of sea water. She is trim and feminine, wears a soft pink nail polish and uses lipstick and powder expertly, but sparingly" (Durgin 1951: 1). At the *New York Herald Tribune* Book and Author Luncheon, Carson observed, "People often seem to be surprised that a woman should have written a book about the sea.... some people are further surprised to find that I am not a tall, oversize, Amazon-type female" (quoted in Lear 1998: 77).

## DEALING WITH FAME

Rachel found it increasingly difficult to maintain her private life. Shortly after the publication of *The Sea Around Us,* while she was on a trip south to start research on *The Edge of the Sea,* she went to a local beauty salon. While sitting under a hair dryer, which Rachel "considered an inviolate sanctuary," the owner abruptly turned it off because an admirer wanted to talk to her (quoted in Brooks 1972: 131). On the same trip, while she was staying in a motel with her mother in Myrtle Beach, a fan seeking an autograph barged into their room while Rachel was still in bed.

Also in 1951, upon returning from a productive summer doing research in Woods Hole and Boothbay Harbor, Rachel found out that her niece Marjorie, who had not been in good health, was several months pregnant after having an affair with a married man. Rachel struggled to keep the situation a secret and maintain her niece's privacy, a difficult task considering the celebrity status she had attained. She found it hard to enjoy her achievements once she knew of her niece's predicament. Rachel's grandnephew, Roger Allen Christie, was born on February 18, 1952. Marie Rodell and Rachel's other friends Dorothy Algire and Alice Mullen were the only people she told at the time. She later wrote, "all that followed the publication of *The Sea*—the acclaim, the excitement on the part of critics and the public at discovering a 'promising' new writer—was simply blotted out for me by the private tragedy that engulfed me at precisely that time. I know it will never happen again, and if ever I am bitter, it is about that" (quoted in Freeman 1995: 148). After Roger's birth, Rachel gradually informed friends that Marjorie had been married for a short time and had had a child.

Due to the unexpected success of *The Sea Around Us*, Rachel Carson had achieved a level of fame that she had never expected and wrote to Marie Rodell, "Heavens, is this all about me—it's really ridiculous!" (quoted in Lear 1997: 201). Even before its publication, she was inundated with requests for speaking engagements and radio and television appearances. In spite of her discomfort with public speaking, she gave many lectures and finally agreed to do one television interview. Preferring book signings to speeches, however, she turned down many offers to speak and at one point told her editor, "enough is enough" (227). She revealed a sense of humility in her acceptance speech for the National Book Award in which she claimed, "If there is poetry in my book about the sea, it is not because I deliberately put it there, but because no one could write truthfully about the sea and leave out the poetry" (quoted in Lear 1998: 91). An unpretentious person who once confessed, "I am always more interested in what I am about to do than in what I have already done," she would much rather have used the time spent as a celebrity working on her next book and did not particularly care for all of the attention given her (quoted in Lear 1997: 287). With this attitude she undertook writing the book that would complete her trilogy of the sea.

*About five years old, reading to her dog Candy. Family photo. Used with the permission of the Rachel Carson Council, Inc.*

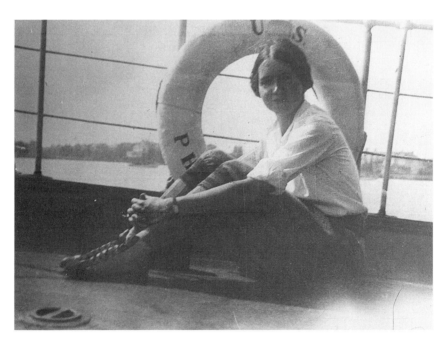

*At Woods Hole, Massachusetts, on a Fish and Wildlife Service vessel (1929). Photo by Mary Frye. Used with the permission of the Rachel Carson Council, Inc.*

U.S. Fish and Wildlife Service portrait
(mid-1940s).

With her dog Rags at her home in Silver
Spring, Maryland (c. 1940). Used with the
permission of the Rachel Carson Council,
Inc.

*Watching migrating birds on Hawk Mountain in eastern Pennsylvania (1945). Photo by Shirley Briggs. Used with the permission of the Rachel Carson Council, Inc.*

On the University of Miami
boat Nauplius (1949). Photo
by Shirley Briggs. Used with
the permission of the Rachel
Carson Council, Inc.

On the dock at Woods Hole,
Massachusetts (1951).
Photo by Edwin Gray. Used
with the permission of the
Rachel Carson Council, Inc.

*Examining specimens through a microscope (1951). Photo by Brooks Studio. Used with the permission of the Rachel Carson Council, Inc.*

*Exploring a tide pool near her Southport Island, Maine, home (1955). Photo by Shirley Briggs. Used with the permission of the Rachel Carson Council, Inc.*

*Working at home (c. 1951). Photo by Shirley Briggs. Used with the permission of the Rachel Carson Council, Inc.*

Above: On the deck of her Southport Island, Maine, cottage (1961). Photo by Bob Hines. Used with the permission of the Rachel Carson Council, Inc.

Right: Portrait (1962). Provided by Yale Collection of American Literature. The Beinecke Rare Book and Manuscript Library, Yale University, New Haven, Connecticut.

# Chapter 7

# COMPLETING THE
# SEA TRILOGY

## WORKING ON *THE EDGE OF THE SEA*

Rachel Carson's next project, the last installment in her trilogy about the sea, would explore the "elusive and indefinable boundary" of the seashore (Carson 1955: 1). The shore had always interested her because the creatures living there had to adapt to the unique environment where land and water met. The environment between ocean and earth had a mysterious quality because of "the interchangeability of land and sea in this marginal world of the shore" (6). It was in a state of constant transition as the tides rolled in and out and the waves beat upon the shore, while at the same time, it also seemed changeless and eternal. Although anyone could go there, human beings were always visitors to this world.

Distracted by all of the attention paid to *The Sea Around Us* and by her family obligations, Rachel was unable to work much on her next book during 1952, even after she resigned from the Fish and Wildlife Service. While she assisted Marjorie with baby Roger and cared for her aging mother, she had little time for writing. According to the contract she had signed with Houghton Mifflin, her seashore guide was due in March 1953, but she was making major revisions to the overall structure of the book that would require a lot more time and work.

Carson intended to write a book that would be considered a sequel to *The Sea Around Us*. Whereas her previous book had been concerned with the physical and geological aspects of the ocean, this one would discuss the biology of the sea. She did not want to simply identify and describe the sea creatures as in a field guide, but rather to explain their complex relationships with one another, their environment, and human beings. The working title, *Guide to Seashore Life on the Atlantic Coast,* was changed to *Rock, Sand, and Coral, a Beachcomber's Guide to the Atlantic Coast,* reflecting her shift in emphasis from a simple categorization of shore creatures to a more complicated ecological analysis of the three major environments in which they lived. Brief descriptions of various creatures were finally relegated to a lengthy appendix.

In the spring of 1952, Rachel did find time to conduct field research for her next book along a number of southern beaches. In the Florida Keys, she met up with her illustrator and friend Bob Hines, staff artist for the Fish and Wildlife Service, and they spent three weeks together as he drew pictures of the creatures she collected with him. Hines, who had replaced Kay Howe after her resignation from the FWS in July 1948, was a midwesterner and former employee of the Ohio Division of Conservation. Although he was unfamiliar with the ocean, he had no problem drawing the creatures that lived there. His pencil drawings were characterized by a lifelike quality because he did not work from preserved specimens but from living animals observed in their natural habitat. After he drew a creature, Rachel was careful to return the specimen to where she had found it.

Rachel considered the illustrations by Bob Hines to be "a substantial part of the book" and "so beautifully and satisfyingly done" that she wanted the book jacket to include a blurb about Hines (quoted in Lear 1997: 270). Following the publication of her book, eventually titled *The Edge of the Sea,* one critic praised Hines's work almost as much as Carson's, calling the book "the product of two naturalists working in close cooperation, each one scientifically trained and each an artist, the one with a pen and the other with a pencil" (Berrill 1955: 30). Hines was equally impressed by Carson and particularly admired her determination. On one occasion, after spending hours exploring some chilly tide pools in Maine, she was so numbed by the water that he had to carry her back to the car. She required similar assistance in a Florida mangrove swamp when, due to her overdedication, she had stayed beyond the point of exhaustion.

## CONSIDERING ANOTHER WRITING PROJECT

Rachel had always been fascinated by the process of evolution, and, although it was usually considered to transpire over thousands or even millions of years, she believed that it could actually be witnessed in the dynamic coastal environment. While exploring the dunes of Saint Simons Island, Georgia, in 1952, she wrote the following passage in her field notebook:

I stood where a new land was being built out of the sea, and I came away deeply moved. Although our intelligence forbids the idea, I believe our deeply rooted attitude toward the creation of the earth and the evolution of living things is a feeling that it all took place in a time infinitely remote. Now I understood. Here, as if for the benefit of my puny human understanding, the processes of creation—of earth building—had been speeded up so that I could trace the change within the life of my own contemporaries. The changes that were going on before my eyes were part and parcel of the same processes that brought the first dry land emerging out of the ancient and primitive ocean; or that led the first living creatures step by step out of the sea into the perilous new world of earth. (quoted in Lear 1998: 131)

In December 1952, while Rachel was struggling to complete *The Edge of the Sea*, she was asked by Dr. Ruth Nanda Anshen to write a book on evolution to be included in the World Perspectives Series published by Harper & Brothers. Despite her other commitments, she signed a contract to write a book tentatively called *Origin of Life* and was sent a $5,000 advance. At first she was excited about this project but then suggested that the topic be broadened to the relation of life to its environment. She began to realize that she would rather write this book for Houghton Mifflin as a separate work rather than for Harper & Brothers as part of a series. But she wanted to fulfill the contract she had signed and agreed to write a short book about the original topic of evolution, with the intention of following this with a longer book on the broader topic she had in mind for Houghton Mifflin.

## SUMMERING ON SOUTHPORT ISLAND

After Rachel returned from her trip south in June 1952, she went to Woods Hole to conduct intensive research for her book in a private lab at the Marine Biological Laboratory. Her mother came along to help out at their rented house. Rachel hired a young researcher and also had assistance from her friend Alice Mullen. During that summer, Rachel spent several weekends in Boothbay Harbor, Maine, where she had dreamed about owning her own cottage by the sea since her first trip there.

Six years earlier, Rachel had celebrated her tenth anniversary at the FWS by taking a month off and rented a small cabin on the Sheepscot River near Boothbay Harbor where she and her mother stayed. She fell in love with the rocky coast of Maine that summer as she watched birds, explored tide pools, and hiked through the surrounding woods and hills. She began to dream of buying her own place there in which to spend summers—a dream that would require more money than she could ever make as a government employee.

By 1952, having become a best-selling author with a much larger income, Rachel began to inquire about homes available in the Boothbay area, but nothing in decent shape seemed affordable or offered the privacy she wanted. Instead, she bought some heavily wooded land on Southport Island, overlooking the estuary of the Sheepscot River, and had a small cottage built there, completed the following summer. Rachel "preferred to live simply, and the cottage reflects her tastes" (Stinnett 1992: 39). In her living room, a large picture window overlooked her private 140-foot shoreline. The Sheepscot River was so deep that whales and seals sometimes swam right past Rachel's house. In this location, with plenty of tide pools to explore, she could conduct a great deal of research for *The Edge of the Sea*. In 1955, she bought additional property next to her original lot, giving her even more shoreline as well as additional privacy.

In July 1953, Rachel went to her new cottage in Maine with her mother and spent time working on her book. The deadline for *The Edge of the Sea* had been extended when it became obvious that Carson would be unable to complete it by the original deadline of March 1953. By this time, her nieces were grown and lived on their own, but Marjorie and her son Roger often came to visit. Shortly after moving in, she met her neighbors on Southport Island, Dorothy and Stanley Freeman, who were fans of Carson's writing. They had written to welcome her to the area after they had read in a local paper that she would be living there.

Although Dorothy was nine years older than Rachel, they immediately sensed that they were kindred spirits who shared a great love of nature and the ocean. They could also commiserate about caring for aging mothers. Dorothy had been coming to her family's Southport Island cottage since she was an infant. Before her marriage, she had worked for the Massachusetts Department of Agriculture in the Cooperative Extension Service and became the first female regional director of the 4-H Club. During the year, Dorothy lived in West Bridgewater, Massachusetts, almost five hundred miles away from Rachel's home in Maryland, so they saw each other infrequently except for their summers together.

Rachel and Dorothy met only briefly a few times that first summer, but they began writing letters to each other that enabled their friendship to develop through the year into "a deep and loving relationship based on shared sympathies, mutual understanding, and unwavering devotion"

(Lear 1997: 248). This relationship would be one of the most important in Rachel's life. In one of her early letters, Rachel wrote, "It seems as though I had known you for years instead of weeks, for time doesn't matter when two people think and feel in the same way about so many things" (quoted in Freeman 1995: 6).

Rachel kept Dorothy informed about her progress on *The Edge of the Sea* because she considered Dorothy to be not only a dear and understanding friend but also "my 'ideal reader'—the kind of person for whom I am writing" (quoted in Freeman 1995: 33). *The Edge of the Sea* was dedicated to Dorothy and Stanley Freeman "who have gone down with me into the low-tide world and have felt its beauty and its mystery" (Carson 1955: dedication page). Finding it hard to write with the distractions created by family responsibilities, she expressed her frustrations to Dorothy, writing, "I think sometimes in regard to the book that if only there were someway, without anyone being hurt by it, that I could be free of all responsibility and worry for even one month...it would make all the difference in the world. But that couldn't be, and it's the feeling that there is no way out that gets me down" (quoted in Freeman 1995: 44–45). The many letters that were exchanged over their twelve-year correspondence contain some of Rachel's most beautiful nature writing, as well as discussions about literature and the writing life. Rachel and Dorothy were concerned that these very personal letters, which provide a great deal of insight into their private lives, might fall into the hands of someone who would misinterpret them, and at times, thought that they should be destroyed. But in 1995, the letters were published in *Always, Rachel*, edited by Dorothy's granddaughter Martha Freeman.

## POLITICAL VIEWS

Rachel Carson had been disappointed by the election of Dwight D. Eisenhower to the presidency in November 1952. She feared that Republican policies regarding conservation would favor big business and threaten the wilderness areas of the country. The administration did indeed eliminate a number of policies regarding conservation that were begun during Franklin D. Roosevelt's presidency. In January 1953, Rachel was particularly disturbed by the appointment of Douglas McKay, a former used-car salesman and businessman from Oregon, as secretary of the interior. McKay, who referred to those interested in conservation as "long-haired punks," sought to open up federal lands to a greater extent to lumber and mining companies and other commercial enterprises (quoted in Brooks 1980a: 250).

Although Carson could not criticize the government while she was a federal employee, once she resigned from her position, she felt free to make her views known. When her former boss, Albert Day, was removed as director of the Fish and Wildlife Service and other professionals were replaced by inexperienced political appointees, she wrote a letter to the editor of the *Washington Post*. She boldly wrote, "It is one of the ironies of our time that, while concentrating on the defense of our country against enemies from without, we should be so heedless of those who would destroy it from within" (Carson 1953: A26). The letter was reprinted in the August 1953 issue of *Reader's Digest*. Her worst fears were realized when the government announced a plan to build a dam that would inevitably flood Dinosaur National Monument and other national monuments in Colorado and Utah.

## FINISHING *THE EDGE OF THE SEA*

In December 1953, Carson participated in an American Association for the Advancement of Science symposium in Boston that focused on the sea. At this meeting, she presented a paper titled "The Edge of the Sea," which was significant because it was "the only purely scientific paper she ever gave to a professional academic organization" (Lear 1997: 250). It was also noteworthy because the title of her paper became the final title of her book. This paper presented some of the ideas that she would discuss at great length in *The Edge of the Sea*, including the evolution of shore creatures and the relationship between the coastal environment and the creatures that inhabit it.

Rachel was determined to finish *The Edge of the Sea* during the summer of 1954. William Shawn of the *New Yorker*, who was instrumental in ensuring the popularity of *The Sea Around Us* when he condensed it in his magazine, was interested in doing the same for *The Edge of the Sea* after reading the chapter "Rim of Sand," sent to him by Marie Rodell. But Rachel had to complete the rest of the book first. The care of her mother, however, demanded an increasing amount of her time, as did the mundane chores that her mother had previously taken care of.

Dorothy Freeman suggested that Rachel hire a housekeeper who could also look after her mother, but Rachel was hesitant. She was grateful that Stan and Dorothy often took Roger on outings with their granddaughter Martha, who was a year younger, so that she could work with fewer interruptions. When she returned to her home in Maryland in October 1954, she did hire a full-time housekeeper because she had promised Paul Brooks that the completed manuscript would be in his hands by January 1955. She finally got most of it to him by March but still had to work on the lengthy appendix as well as the prologue, epilogue, and captions for the illustrations.

## AN OVERVIEW OF *THE EDGE OF THE SEA*

In the preface to *The Edge of the Sea,* Carson explained, "To understand the shore, it is not enough to catalogue its life. Understanding comes only when, standing on a beach, we can sense the long rhythms of earth and sea that sculpted its land forms and produced the rock and sand of which it is composed; when we can sense with the eye and ear of the mind the surge of life beating always at its shores—blindly, inexorably pressing for a foothold" (Carson 1955: vii). She stated that her purpose in writing the book was "to interpret the shore in terms of that essential unity that binds life to earth" (viii). Later, she asserted, "Nowhere on the shore is the relation of a creature to its surroundings a matter of single cause and effect; each living thing is bound to its world by many threads, weaving the intricate design of the fabric of life" (14). These quotes reveal an ecological perspective that was revolutionary for its time, especially among the general audience for whom the book was intended.

In chapter 1, "The Marginal World," which introduces the shore environment, Carson wrote:

> The shore is an ancient world, for as long as there has been an earth and sea there has been this place of the meeting of land and water. Yet it is a world that keeps alive the sense of continuing creation and of the relentless drive of life. Each time that I enter it, I gain some new awareness of its beauty and its deeper meanings, sensing that intricate fabric of life by which one creature is linked with another, and each with its surroundings. (Carson 1955: 2)

She described the shore as a place "where the drama of life played its first scene on earth and perhaps even its prelude; where the forces of evolution are at work today, as they have been since the appearance of what we know as life" (7). The next chapter, "Patterns of Shore Life," describes the evolution of sea creatures as revealed in fossil records, explaining how the basic ocean forces of waves, currents, and tides effect this evolutionary process as the momentum of waves, the temperature of the water, and the time that an area is underwater all contribute to the creation of distinct life zones.

The rest of *The Edge of the Sea* is divided into three parts, each focusing on a particular type of shore environment, rather than being organized like a typical field guide that classifies creatures based on their physical appearance. Because equality is given to all creatures, they are blended together into a single habitat. "The Rocky Shores" covers the rocky coastline starting at Cape Cod and going north where life is predominantly influenced by the tides, while "The Rim of Sand" discusses the sandy beaches from Cape Cod south where waves are the strongest force on living creatures. Finally, "The Coral Coast" focuses on the coral reefs and mangrove forests of the Florida Keys where the ocean currents most greatly affect life.

Most of the examples and anecdotes in chapter 3, "The Rocky Shores," are based on Carson's observations of the Maine coast, particularly that portion right outside her own door on Southport Island. In geological terms, this type of shoreline is relatively new in the history of the earth, as its jagged rocks have not yet been pounded into sand by the surf, and the evergreen forest extends right to the edge of the land. The various creatures that inhabit this environment are intricately interrelated, as this chapter explains. For example, as limpets feed on algae covering the rocks, the rocks become slippery, making it easier for barnacle larvae to attach to them. When the barnacles die, their shells remain to provide shelter for other small creatures of the shore such as baby periwinkles, insects, or anemones.

Chapter 4, "The Rim of Sand," begins with a geological history of sandy beaches and explains how their origins differ from the rocky coast. The sandy coastline is older than the rocky shore because most sand is simply eroded rock. Sand, which is constantly in motion, is actually stronger than rock and is almost indestructible. The creatures of the sandy beach are mostly hidden below the surface, seeking protection from the fish that come in with the tide and the birds that come at low tide. Although Carson wrote that "the beach has a lifeless look, as though not only uninhabited but indeed uninhabitable," she found much life in this apparently lifeless zone (Carson 1955: 131). Creatures such as the ghost crab and the sand hopper, for instance, demonstrate the process of evolution in action as they appear to be making the transition from sea organisms to inhabitants of the land.

The coral coast of the Florida Keys that is described in chapter 5 is an area that contrasts markedly with the previous two environments and is unique in the world because living things themselves form the coral reef. This chapter also focuses on the environment of mangrove swamps that are formed in sheltered bay areas when the branches of mangrove trees spread out and root themselves, creating more tree trunks that grow in the water and form swamps and islands.

A brief concluding chapter, "The Enduring Sea," reflects a recurrent theme in Carson's work—the ocean's timelessness. While the coast is always changing, she believed that the sea remains eternal: "Once this rocky coast beneath me was a plain of sand; then the sea rose and found a new shore line. And again in some shadowy future the surf will have ground these rocks to sand and will have returned the coast to its earlier state" (Carson 1955: 249). She also observed, "On all these shores there are echoes of past and future: of the flow of time, obliterating yet containing all that has gone before; of the sea's eternal rhythms" (250). Because *The Edge of the Sea* has no formal structure aside from the three major categories of coastline and is not organized in a traditional way, a lengthy appendix provides a more rigid organization and is included "for the convenience of those who like to pigeonhole their findings neatly in the classification schemes the human mind has devised" (viii).

Although Carson's writing in *The Edge of the Sea* often has an objective tone, as in *The Sea Around Us,* she occasionally inserted a first-person point of view by including a personal anecdote or an observation of something she had seen through her microscope or during her explorations of the tide pools on her property and elsewhere. For example, she wrote, "One of my own favorite approaches to a rocky seacoast is by a rough path through an evergreen forest that has its own particular enchantment" (Carson 1955: 41). Later, she recounted, "Curious about the early stages of this abundant snail, I have gone down into my own rockweed forests on the summer low tides to search for them" (84).

After Rachel Carson completed the manuscript for *The Edge of the Sea,* she spent a summer on Southport Island free from the pressure of having a book to write. Her domestic situation, however, enervated her as she cared for her aging mother, as well as Marjorie, whose diabetes had worsened, and Roger. She complained to Dorothy, "When I feel, as I do now, the pressure of all the things that seem worth doing in the years that are left, it seems so silly to be spending my time being a nurse and housemaid" (quoted in Freeman 1995: 151). As a best-selling author, she was inundated with offers of writing projects, and, despite her exhaustion, she soon began to consider what she would do next.

# Chapter 8

# REACHING NEW AUDIENCES

## REACTION TO *THE EDGE OF THE SEA*

While Rachel was in Maine, the *New Yorker* published the condensed version of *The Edge of the Sea* in two parts on August 20 and 27, 1955. The publication was met with praise, but the fan mail was much less voluminous than it had been for *The Sea Around Us*. Rachel Carson was no longer a new author, so the excitement over this book was less intense. Rachel wrote to Dorothy, "I know that, even if this book achieves acceptance, acclaim, and sales that by any reasonable standards amount to 'success'—still, by comparison with *The Sea*, it will fail. What I want for *The Edge of the Sea* is for it to be judged on its own merits, but that is most unlikely to happen" (quoted in Freeman 1995: 127).

*The Edge of the Sea* was published on October 26, 1955. A month later, it was eighth on the *New York Times* best-seller list. Considering that the book was limited in its scope to the eastern coast of the United States and might not be of interest to readers nationwide, it did remarkably well, rising finally to third place on the *Times* list and staying there for five months. *The Edge of the Sea* also faced competition from another book about the ocean, *Gift from the Sea* by Anne Morrow Lindbergh, which topped most best-seller lists for over a year. The public often mistook Carson and Lindbergh, as well as their books.

The *New York Times* review on publication day stated that Carson "has done it again" and that she "can do no wrong," calling the book "as wise and wonderful" as *The Sea Around Us* (Poore 1955: 29). The *Saturday Review* described it as an "entrancing work" that was also "permeated by sound biological science" (Berrill 1955: 30). Harry Ellis commented, "*The Edge of the Sea* is pitched, perhaps, in a lesser key than was *The Sea Around Us,* if only because the intertidal world is a more limited subject than was the whole sea itself. In her new book, however, Miss Carson's pen is as poetic as ever, and the knowledge she imparts is profound. *The Edge of the Sea* finds a worthy place beside Miss Carson's masterpiece of 1951" (Ellis 1955: 8B). Critic Jacquetta Hawkes noted, "Miss Carson succeeds admirably in conveying a sense of the richness and intricate interrelatedness of the life she describes. No jeweller could create a work so delicately interlocked and encrusted" (Hawkes 1956: 17).

Carson won a number of honors and awards following the publication of *The Edge of the Sea.* The Museum of Science in Boston elected her an honorary fellow in 1955. The following year, the National Council of Women gave her book a citation for "outstanding book of the year." She won the Achievement Award of the American Association of University Women, which came with a $2,500 cash prize. Her alma mater, the Pennsylvania College for Women, which had recently renamed itself Chatham College, gave her a Distinguished Service Alumnae Award. More important than any awards, however, Carson "had become an authority whose opinions were quoted, rather than an unknown writer who quoted authorities" (Glotfelty 1996: 158).

## "SOMETHING ABOUT THE SKY"

Rachel wanted to take a break from book writing and become involved in projects that might reach new audiences. She was asked to write a script for the CBS television documentary series *Omnibus*. Although she had little interest in television, she accepted the invitation because she felt that she could reach a larger audience through this new medium. An eight-year-old who was a regular viewer of the series wrote a letter to *Omnibus* requesting that a program be produced "on something about the sky" (quoted in Lear 1997: 280). The producers suggested this topic to Carson, although they would probably have done a show on any subject that the best-selling author selected. The episode "Something about the Sky" aired on March 11, 1956.

Rachel had always been interested in clouds and had even considered writing a book on that topic, which she would title *The Air Around Us.* From her perspective, clouds were intricately related to the sea that she loved because rain falling from the clouds continually replenished the seas. In "Something about the Sky," she wrote about the air in terms that could also be used to speak about the ocean, describing clouds as being adrift and calling wind "atmospheric waves" (quoted in Lear 1998: 178). She considered the sky to be as timeless and eternal as the sea: "The clouds are as old as the earth itself—as much a part of our world as land or sea. They are the writing of the wind on the sky. They carry the signature of the masses of air advancing toward us.... Most of all they are cosmic symbols of a process without which life itself could not exist on earth" (176). Rachel was so pleased with "Something about the Sky" that she finally bought a television set of her own.

## "HELP YOUR CHILD TO WONDER"

Another project that Rachel completed at this time was an article for *Woman's Home Companion* titled "Help Your Child to Wonder," published in July 1956. This autobiographical article about instilling in children an appreciation of the natural world was based on her own experiences with Roger as she tried to teach him about nature. A private person, Rachel seldom divulged details of her personal life in her work. "Help Your Child to Wonder," however, revealed a side of Rachel that had never been detected in her other writings. "Help Your Child to Wonder" was reprinted in *Reader's Digest*, and both Marie Rodell and Paul Brooks suggested that Rachel expand the article into a book. Although she never got around to completing this project, after her death the article was reprinted with numerous photographs taken primarily by Charles Pratt in a book titled *The Sense of Wonder*.

Rachel began the article by writing about the autumn night when she wrapped baby Roger in a blanket and took him down to the beach to introduce him to the ocean. She also recounted her rainy day walks through the woods with her nephew. She emphasized the importance of a child having at least one adult with whom to share the beauty of nature so that the "sense of wonder" with which every child is born does not fade as adulthood approaches. She asserted:

> A child's world is fresh and new and beautiful, full of wonder and excitement. It is our misfortune that for most of us that clear-eyed vision, that true instinct for what is beautiful and awe-inspiring, is dimmed and even lost before we reach adulthood. If I had influence with the good fairy who is supposed to preside over the christening of all children I should ask that her gift to each child in the world be a sense of wonder so indestructible that it would last throughout life, as an unfailing antidote against the boredom and disenchantments of later

years, the sterile preoccupation with things that are artificial, the alienation from the sources of our strength. (Carson 1956: 46)

In "Help Your Child to Wonder," Carson suggested early morning walks to hear the birds in spring and nighttime outings to look up at the stars. A scientific knowledge of nature is unnecessary to help a child to experience this "sense of wonder"; all that is required is a love of nature. She wrote, "Exploring nature with your child is largely a matter of becoming receptive to what lies around you. It is learning again to use your eyes, ears, nostrils and fingertips, opening up the disused channels of sensory impression" (Carson 1956: 47). It is not essential to know the different songs of the birds or the names of the trees to appreciate these things. In fact, she felt that the ability to identify the things of nature is useless as a means in itself: "It is possible to compile extensive lists of creatures seen and identified without ever once having caught a breath-taking glimpse of the wonder of life" (48). As in *The Edge of the Sea*, Carson discounted classification schemes such as those used in field guides, considering it more essential to understand the relationships among the creatures of the natural world. After remembering the famous oceanographer Otto Pettersson, she wrote, "The lasting pleasures of contact with the natural world are not reserved for such scientists but are available to anyone who will place himself under the influence of earth, sea and sky and their amazing life" (48).

## "THE LOST WOODS"

Just as Rachel felt that children could benefit from exposure to the wonders of nature, so she believed that all people are enriched by visits to the forest, seashores, and other places of beauty. Slowly these places were disappearing because of development, however, and the need for preservation was becoming more urgent. Rachel had always wanted to help save such places. She had written to Dorothy Freeman about "the general problem that is so close to my heart—the saving of unspoiled, natural areas from senseless destruction" (quoted in Freeman 1995: 16).

Back in 1950, upon completion of her manuscript for *The Sea Around Us*, Rachel had visited Island Beach, New Jersey, located on the southern tip of Barnegat Peninsula. This privately owned area, which was "one of the last remaining examples of pristine barrier ecology on the mid-Atlantic coast" and provided a preserve for a wide variety of birds, had been put up for sale (Lear 1997: 178). The Island Beach National Monu-

ment Committee, headed by Rachel's friend Richard Pough, was organized to save the area from development. Rachel went to Island Beach
with the hope of writing an article about it that would help the conservation efforts. Although she was unable to complete the necessary research
because of car problems and bad weather, her visit there was an early indication of her support for saving such areas.

In a speech that Carson delivered in 1954 at the Matrix Table Dinner
of Theta Sigma Phi, a sorority of almost a thousand women journalists,
she shared her thoughts on the value of the natural world and the urgency
to save it. In her speech, "The Real World around Us," she asserted, "I believe that whenever we destroy beauty, or whenever we substitute something man-made and artificial for a natural feature of the earth, we have
retarded some part of man's spiritual growth" (quoted in Lear 1998: 160).
She further stated:

> Mankind has gone far into an artificial world of his own cre-
> ation. He has sought to insulate himself, with steel and con-
> crete, from the realities of earth and water. Perhaps he is
> intoxicated with his own power, as he goes farther and farther
> into experiments for the destruction of himself and his world.
> For this unhappy trend there is no single remedy—no panacea.
> But I believe that the more clearly we can focus our attention
> on the wonders and realities of the universe about us, the less
> taste we shall have for destruction. (163)

As examples of the "artificial world" that man had created she mentioned
the Levittown, Pennsylvania, suburbs where thousands of identical
houses were built on clear-cut land. In her speech, she also protested a
proposal to build a major highway through one of her favorite places,
Rock Creek Park in Washington, D.C.

After Rachel began spending summers in Maine, she became concerned about an area of forest and shoreline between her cottage and
Dorothy's house that had always been a special place for them. As they
began to notice that all around them the population was growing and land
was being bought, they began to dream they could buy the land and preserve it from future development so that it could forever be enjoyed as a
sanctuary of nature. They referred to this place as the "Lost Woods" after
the title of an essay by one of their favorite authors, H. M. Tomlinson.
Rachel described it in a letter to her friend Curtis Bok, President Judge of
the Pennsylvania Supreme Court, writing that the "charm" of the Lost
Woods "lies in its combination of rugged shore rising in rather steep cliffs

for the most part, and cut in several places by deep chasms where the storm surf must create a magnificent scene." She further wrote, "Behind this is the wonderful, deep, dark woodland—a cathedral of stillness and peace." Finally, she asserted, "It is a treasure of a place to which I have lost my heart completely" (quoted in Lear 1998: 173–74). At a time when Rachel was feeling particularly frustrated by her inability to accomplish much writing because of her personal responsibilities, her involvement in this project to save the Lost Woods gave her the sense of purpose that she so desperately needed.

Carson also became involved in organized conservation efforts. She was instrumental in the formation of a Maine chapter of the Nature Conservancy and became honorary chairman as the fourth field office of the organization opened on November 26, 1956. She was also invited to serve on the board of directors of the American Foundation established by Edward W. Bok, father of Judge Curtis Bok, which supported the Mountain Lake Sanctuary in Lake Wales, Florida. Through her involvement in these causes, she began to formulate her own plan to save the place that was so special to her—the Lost Woods.

Rachel knew that she had to raise a large sum of money to buy the Lost Woods. Most of the land was owned by Gustav Tenggren, a children's book illustrator. Although Rachel was a best-selling author, she still did not have enough to cover the cost of the land while continuing to maintain her lifestyle. She had decided to build an addition on her Maine cottage to accommodate Marjorie and Roger and also planned to have a new home built in Silver Spring, Maryland.

Rachel considered some potentially lucrative writing projects. Simon & Schuster had proposed to Rachel that she write a children's edition of *The Sea Around Us* to be included in its Golden Book series, but she had refused this offer because she simply didn't have time to do it herself and would not allow someone else to do it. She reconsidered after she realized that the royalties from this book could support Marjorie and Roger. The contract drawn up by Marie Rodell stated that the money Rachel earned from the book would go to her niece and that the publisher would hire a writer but Rachel would have final approval of the manuscript. She was also asked by Simon & Schuster to reconsider an offer to edit an anthology of nature writing that would have a print run of one hundred thousand sets at $20 per set. The royalties from this project could go toward purchasing the Lost Woods. She also compiled a bibliography of recommended biology books and an accompanying essay that was published in the 1956 edition of the reference source *Good Reading*. She was well paid for her effort, but the project was more complicated than she had expected. In addition to these undertakings, she began to explore the possibility of a documentary based on *The Edge of the Sea* even though she had had such a bad experience with the film version of *The Sea Around Us*.

## THE DEATH OF MARJORIE

Rachel was feeling very positive as 1956 drew to a close. She wrote to Dorothy:

> And now, like the scattered parts of a puzzle suddenly falling into place, everything seems possible! When I stop to pinch myself I can scarcely believe it. It's like an answer to prayer— and yet I confess I have not prayed for it, unless my life, not words, were the prayer. But for a good many years I have believed that in order to achieve one must dream greatly—one must not be afraid to think large thoughts. And now, suddenly, as though it was "meant to be," the way seems to be opening up. (quoted in Freeman 1995: 202)

Unfortunately, Rachel's optimism was to be short-lived. In January 1957, Marjorie's health deteriorated further when she was hospitalized for pneumonia. Although she returned home briefly and Rachel believed she would recover, on January 30, at the age of thirty-one, Marjorie passed away, leaving her son Roger orphaned. Marjorie's sister Virginia and Rachel's brother Robert, neither of whom was as close to Roger as Rachel, had no interest in taking in the child. Rachel realized that the only solution was to adopt Roger herself, and at the age of fifty, she became the parent of a highly energetic five-year-old who constantly demanded her attention as she continued to care for her eighty-eight-year-old mother.

Rachel confided in Dorothy, "For I grow more conscious with each passing week that my life will never again be the same, and that when it might otherwise be possible to do the things I had thought to do, the sands will have run too low" (quoted in Freeman 1995: 219). To further erode her mood, her good friend Alice Mullen died suddenly in April. It was almost impossible for Rachel to focus on her work on the anthology or anything else except some editing of the juvenile edition of *The Sea Around Us*. Rachel, her mother, and Roger moved into their newly constructed Silver Spring home in July 1957, and after settling in a bit, left for the renovated cottage in Maine in August for a short stay. Rachel was unhappy because she spent most of her time taking care of Roger and her mother. She was also discouraged when Gustav Tenggren told her that he was not interested in selling his land at that time, and even if he were, the price was much more than Rachel had anticipated. She began to feel powerless to save the Lost Woods.

## "OUR EVER-CHANGING SHORE"

At this time, Carson accepted an offer from *Holiday* magazine to write an article about the shore, hoping to use the opportunity to advocate the conservation of coastal areas. She reasoned that she would be accomplishing her purpose of preserving the environment in a different way. It was a supreme effort to write the article because she was constantly distracted by the demands of raising Roger. "Our Ever-Changing Shore" was published in a special issue of *Holiday* in July 1958 devoted to "Nature's America." It included descriptions of some of Carson's favorite beaches from Maine to Florida and then emphasized how these beautiful places, each one unique and constantly changing, were disappearing because of development.

In this article, Carson once again asserted the geological perspective in which the influence of man has been only a very recent development in the millions and millions of years that the shores have existed, equating a "human generation" with "a mere second in earth history" (Carson 1958: 120). Believing that human beings, as always, exist on the periphery of this environment, she wrote, "For the ocean has nothing to do with humanity. It is supremely unaware of man, and when we carry too many of the trappings of human existence with us to the threshold of the sea world our ears are dulled and we do not hear the accents of sublimity in which it speaks" (117).

Carson believed that by contemplating the sea from the shore, we might be able to gain spiritual qualities such as "strength and serenity and endurance" (Carson 1958: 120). In her article, she lamented, "The shore might seem beyond the power of man to change, to corrupt. But this is not so. Unhappily, some of the places of which I have written no longer remain wild and unspoiled" (119). Then she asserted, "Somewhere we should know what was nature's way; we should know what the earth would have been had not man interfered.... For there remains, in this space-age universe, the possibility that man's way is not always best" (120). The article ends with a call to preserve the coastline by supporting the National Park Service in its efforts to maintain public areas. Lear wrote, "Her plea for the preservation of the nation's seashores remains one of the most eloquent in contemporary nature writing" (Lear 1998: 113).

Meanwhile, Rachel continued to think about her commitment to write *Origin of Life* for Harper & Brothers, but she worked on it very little during this period and wrote to Dorothy Freeman in February 1958 that she had been "mentally blocked for a long time" (quoted in Freeman 1995: 248). Her perspective on the project was evolving as she began to realize the consequences that the atomic age was having on the earth. She went on to tell Dorothy, "It was pleasant to believe...that much of Nature was forever beyond the tampering reach of man.... It was comforting to suppose that the stream of life would flow on through time in whatever course that God had appointed for it" (248–49). Rachel was becoming increasingly aware that human beings were threatening the natural world that she loved. Not only were they polluting the air and the ocean and destroying the shoreline, by disrupting the delicate balance of nature they were, as Carson would find out, killing the beautiful creatures that inhabited these places.

# Chapter 9

# THE STRUGGLE TO WRITE
## *SILENT SPRING*

Rachel Carson had been concerned about the dangers of pesticides, especially DDT, since her days at the Fish and Wildlife Service. She had always felt that DDT was not the miraculous substance that scientists during World War II had proclaimed it to be and that further research was necessary to determine the true dangers of the pesticide. Her efforts to get an article published on the topic failed, including her rejected proposal to *Reader's Digest* in 1945. The more she learned about the problem, "the more appalled I became. I realized that here was the material for a book. What I discovered was that everything which meant most to me as a naturalist was being threatened, and that nothing I could do would be more important" (quoted in Brooks 1972: 233).

Despite Rachel's strong reservations about pesticide misuse, she did not feel that she had time to write an entire book. In addition to the other projects that she needed to work on, taking care of Roger and her mother were, as always, time-consuming responsibilities leaving her little time for writing. Because she considered the issue so important, however, she tried to interest other authors in writing such a book, but it soon became apparent that she was the most qualified person to do it. Although she thought the book would take, at most, a year to write, as she delved more deeply into the subject of pesticides, she realized that the problem was far worse than she could have ever imagined. The short book that she had planned to write in one year turned into a four-year struggle to complete her most famous work and one of the most influential books of the twentieth century, *Silent Spring*.

# THE HISTORY OF DDT

DDT (dichloro-diphenyl-trichloro-ethane) was first developed in 1874 by a German graduate student who was unable to find a purpose for it. In 1939, Swiss chemist Paul Müller discovered its use as an insecticide while working for Geigy, a chemical company that wanted to find a way to mothproof wool clothing. When Müller discovered that it was a very potent pesticide, farmers in Switzerland began using it to protect their crops from insects. It was also used during World War II by the United States to kill lice and mosquitoes, thus preventing outbreaks of typhus and malaria. Because of DDT, "World War II is thought to be the first major war in which more people died from enemy action than from disease" (Glotfelty 2000: 158). Müller was awarded a Nobel Prize in medicine and physiology in 1948 for his work with DDT.

After World War II, DDT and more than two hundred other pesticides were manufactured by chemical companies in the United States for widespread use by the government, farmers, foresters, and amateur gardeners. Pesticides were popular in postwar America where citizens had "the profound wish for order and control that followed the chaos of that war" (Boucher 1987: 38). In the growing suburbs of the 1950s, people wanted to rid their neighborhoods of the unpleasantness of insects. Pesticide usage grew from 125 million pounds in 1945 to 600 million pounds ten years later.

Pesticides such as DDT appeared to be lethal to a wide variety of insects while being relatively harmless to mammals. The government and chemical companies extolled the benefits of DDT, and public health departments demonstrated its safety by spraying children while eating or playing outside. It was inexpensive and, because it was both a persistent pesticide that did not break down easily in the environment and an insoluble substance that was not washed away by rain, it did not have to be reapplied very often. But there were some scientists who were concerned about the long-term side effects of DDT and considered it, in the words of nature writer Edwin Way Teale, a "two-edged sword" (quoted in Lear 1997: 119).

# THE GROWING PESTICIDE CONTROVERSY

Several situations convinced Carson that a book on the dangers of pesticides would be vitally important. Rachel's friend, ornithologist Robert Cushman Murphy, had become involved in a lawsuit against New York State and the U.S. government for spraying DDT to eradicate gypsy moths, tent caterpillars, and mosquitoes in the area where he lived on Long Island. The pilots of the spray planes were paid by the gallon rather than by the acre and so were motivated to spray more than necessary, drenching areas with pesticides. The threat of gypsy moths infesting the New York City area was given as a reason for the spraying despite the fact that the gypsy moth's habitat is the forest, not urban areas. Cushman was enlisted in the lawsuit by his neighbor Marjorie Spock, whose organic garden was ruined by the spraying. Spock began sending Carson numerous newspaper articles regarding the 1957 trial and the effects of DDT on birds. Marjorie, sister of the famous pediatrician Benjamin Spock, soon became a good friend who visited Carson in Maine and kept her thoroughly informed about developments in the case. Murphy managed to get several other prominent Long Island residents affected by the incident to join him and Marjorie as plaintiffs. The case was eventually considered by the U.S. Supreme Court where it was declined because of a technicality.

Also during that year, the U.S. Department of Agriculture (USDA) began the "fire ant program" to exterminate a type of South American insect that had migrated to the southern United States but was only a minor nuisance. Thousands of acres in the south and southwest were sprayed with chlorinated hydrocarbons such as dieldrin, forty times more toxic than DDT, and heptachlor. Frank Graham, the author of *Since "Silent Spring,"* believed that the USDA considered chemical pesticides to be "a fascinating new toy, which it was arrogantly flaunting at every opportunity" (Graham 1970: 27).

There was great public opposition to the "fire ant program" because of the alleged long-term toxicity of the substances used. In some areas that were sprayed, it was reported that all wildlife was completely destroyed, but the USDA claimed these reports were exaggerated. Another negative effect was an increase in the population of insects that destroyed sugarcane, and in Florida, the spraying resulted in more fire ants than when the program began. In March 1959, Rachel previewed the public service film *Fire Ant on Trial*, produced by the USDA to gain support for the program. She called it "flagrant propaganda in support of a program that has been widely challenged as ill-conceived, irresponsible, and dangerous" (quoted in Lear 1997: 343).

To add to the controversy of pesticide use, Olga Owens Huckins wrote a letter to the *Boston Herald* in January 1958 that detailed the devastation of her backyard bird sanctuary after an episode of aerial spraying of DDT to kill mosquitoes in Plymouth County, Massachusetts, near Cape Cod. Although the spraying was state-sponsored, the residents had not granted permission for it. Huckins had become acquainted with Carson in 1951 when, as literary editor for the *Boston Post,* she wrote a positive review of *The Sea Around Us.* Concerning the devastation caused by the DDT spraying, Huckins wrote in her letter to the *Herald,* "All of these birds died horribly, and in the same way. Their bills were gaping open, and their splayed claws were drawn up to their breasts in agony" (quoted in Brooks 1972: 232).

While the mosquito problem seemed to be worse than ever, Huckins observed that there were no longer such innocuous insects as grasshoppers or beneficial ones like bees in the area where the DDT was sprayed. In her letter to the editor, she demanded that "the spraying of poisons" be stopped until thorough research into the effects of these substances on wildlife and humans could be conducted. She concluded by writing, "Air spraying where it is not needed or wanted is inhuman, undemocratic, and probably unconstitutional. For those of us who stand helplessly on the tortured earth, it is intolerable" (quoted in Brooks 1972: 232). Although she was told that the DDT was harmless, when Huckins learned that it was to be sprayed again since the mosquitoes had not been eradicated, she sent a copy of her letter to her friend Carson and asked for her help in the matter. She thought Carson might know someone in Washington who would have some influence in stopping the spraying of DDT.

## AN ARTICLE BECOMES A BOOK

Marie Rodell continued her search for a magazine that would be interested in a piece written by Carson about the pesticide problem, but this proved difficult. Magazine publishers were concerned about advertisers pulling their ads if such a controversial article were to appear. *Reader's Digest,* however, which had earlier rejected Carson's article about DDT, published editor Robert S. Strother's article "Backfire in the War against Insects" in June 1959. This article criticizing pesticide use appeared after *Reader's Digest* had considered publishing an article in favor of aerial spraying the year before. *New Yorker* editor William Shawn, who had published excerpts from Carson's books, expressed interest in having Carson write an article about pesticides after staff writer E. B. White recommended her writing on this topic.

As Rachel's research revealed the severity of the problem, she reconsidered the possibility of at least contributing to a book on the pesticide issue. She told Paul Brooks that the magazine article she was writing for the *New Yorker* could ultimately serve as a chapter for a book. She might also be able to write an introduction to such a book and help with editing, but that was all she felt able to undertake at the time. Never intending to write the entire thing herself, she signed a contract with Houghton Mifflin in May 1958 for a book with the working title *Control of Nature*. Edwin Diamond, the science editor for *Newsweek*, was hired by Houghton Mifflin to collaborate with her. Because she could not spend a lot of time on the project, she offered to write the beginning and end of the book while Diamond would take care of the middle. She planned to have the manuscript for a short book completed by July to be published in January of the following year.

Although Diamond was at first excited about the opportunity to work with Carson, the literary partnership was a fiasco. The ambitious Diamond and the reserved Carson were completely different personalities who simply did not work well as a team. Diamond was contracted to be coauthor, but Rachel realized that what she really needed was research assistance for which Diamond considered himself overqualified. Although personal problems such as the care of Roger and her mother and her own health problems made it impossible to meet the original deadline, she decided not to collaborate on the project.

As Rachel delved into the subject of pesticides more deeply, the enormity of the problem became more apparent, and she realized this would not be the short book she had anticipated. She became determined to see the project through, writing to Dorothy, who was not very enthusiastic about Rachel's choice of topic, "there would be no future peace for me if I kept silent" (quoted in Freeman 1995: 259). To her former FWS boss Clarence Cottam she wrote, "This was something I had not expected to do, but facts that came to my attention last winter disturbed me so deeply that I made the decision to postpone all other commitments and devote myself to what I consider a tremendously important problem" (quoted in Brooks 1972: 248).

## RESEARCHING *SILENT SPRING*

Because Rachel needed a research assistant, she hired Bette Haney, the daughter of an employee of the Fish and Wildlife Service. A student at Bryn Mawr College who was majoring in biology, the young woman reminded Rachel of herself at that age. Although Haney did not possess the patience her employer had, she came to appreciate Carson's persistence, comparing her to the tortoise in the fable of the tortoise and the hare. Haney thought that Rachel would never finish the book, but she was eventually quite impressed at the amount of work that Carson accomplished at her painfully slow pace. Haney once commented, "As a child of my culture, I had not yet learned to associate progress with that pace" (quoted in Lear 1997: 370).

To write her book, Carson needed information, and because of her many years of government service, she had developed a network of scientists and librarians, as well as staff at organizations such as the Smithsonian Institution and the Audubon Society, who could help her with her research. Her friend Dorothy Algire had become a librarian at the National Institute of Health and was able to procure the numerous obscure documents that Rachel required. Harold Peters, a former FWS biologist, who went to work for the Audubon Society as a research biologist, supplied Carson with accurate and invaluable statistics and information regarding the effects of the government's pesticide use.

Carson consulted many experts including Dr. C. J. Briejèr of the Plant Protection Service in the Netherlands. She also depended on information from Clarence Cottam, who had become director of the Welder Wildlife Foundation in Sinton, Texas, and an influential leader in national policy making regarding conservation issues. Upon completing each chapter after meticulous research, she would send copies to various experts so that they could check the information for accuracy, but they seldom recommended any substantial changes. Clarence Tarzwell, chief aquatic biologist for the Public Health Service, read "Rivers of Death," and her important chapters on cancer were sent to Dr. John J. Biesele, a research scientist at the Sloan Kettering Institute. Cottam reviewed her chapters about wildlife.

Robert Rudd, a zoologist at the University of California at Davis who was also writing a book about pesticides, corresponded with Carson and visited her at her Southport Island cottage. She wrote to him, "I learned long ago that it doesn't matter how many people write about the same thing; each will make his own contribution" (quoted in Lear 1997: 330). Rudd's book, *Pesticides and the Living Landscape,* published in 1964, was described as "a sequel to, and vindication of, *Silent Spring*" (Graham 1970: 176).

During the summer of 1961, Rachel received a letter from Ruth Scott, a naturalist from Pittsburgh who was interested in informing people about the dangers of pesticides. She was coming to Maine with her husband to visit an Audubon Society camp and wanted to meet the famous author. Rachel invited them to her cottage, and the two women realized that they shared many of the same views on nature and environmental issues. From then on, they corresponded often. Ruth was involved in efforts to prevent DDT spraying in an area of Pennsylvania near Rachel's hometown of Springdale. Because Ruth had many important contacts in conservation organizations, she provided Carson with a great deal of assistance in her research.

Not everyone Carson knew was helpful; in fact, some were utterly antagonistic. The Department of the Interior considered her to be a subversive individual. The Department of Agriculture restricted Rachel's access to material at the USDA Agricultural Research Service as her opposition to the fire ant eradication program was revealed and the purpose of her research became known. Rachel had one ally at the USDA, entomologist Reece Sailer, the assistant chief of insect identification and parasite introduction. Although he wrote her informative letters providing information that she needed, he did not want to be revealed as a source. Carson did all she could to protect his identity, and his correspondence never appeared in her collection of personal papers.

## THE DEATH OF MARIA CARSON

In December 1958, Rachel was devastated by the death of her mother following a stroke and pneumonia. Rachel eulogized the woman who was such a major influence on her life in a letter to her friend Marjorie Spock: "Her love of life and of all living things was her outstanding quality, of which everyone speaks. More than anyone else I know, she embodied Albert Schweitzer's 'reverence for life.' And while gentle and compassionate, she could fight fiercely against anything she believed wrong, as in our present Crusade! Knowing how she felt about that will help me to return to it soon, and to carry it through to completion" (quoted in Lear 1997: 338).

Although greatly saddened by her mother's death, Rachel found that the memory of Maria Carson provided her with the inspiration she needed to continue her book. She was also liberated from the responsibility of caring for an aging parent, a responsibility that had become physically and emotionally draining. Mrs. Carson's influence over her daughter had always been powerful. Paul Brooks believed that Rachel had remained single because of her mother, writing "it is probably an understatement to say that Maria Carson never urged Rachel to marry" (Brooks 1972: 242). Although this might have been true, it is certainly possible that Rachel "chose not to marry in order that her creative life might be more fulfilling" (Lear 2000: 211). Whatever the circumstances, the family responsibilities that Rachel had taken on at an early age, together with her ambitious career goals and her own medical problems, had left her little time for much else in her life.

Maria Carson's death set Rachel even further behind schedule on her book, but she agreed with Paul Brooks on completing her manuscript by October 1959 so that it could be published in February 1960. A month before her deadline, with a considerable amount of work remaining on the book, Rachel hired an assistant named Jeanne Davis to replace Bette Haney, who was starting medical school. In addition to possessing the knowledge that the job demanded, Davis was a highly organized individual and just the person that Rachel needed to help her with both research and administrative details. An admirer of Carson's work, Davis was in her early forties and had a degree in economics from the University of California at Berkeley as well as a secretarial degree from Simmons College in Boston. She had worked as a research assistant and editor for two professors at Harvard Medical School while her husband was a medical resident. She described Rachel as "a feminist before her time" who "didn't let anyone push her around" (quoted in "Rachel Carson's Silent Spring" 1993).

## RAISING PUBLIC AWARENESS

Rachel preferred to keep the subject of her book a secret from the general public, but there were several opportunities when she could not remain silent on the subject of pesticide use. In April 1959, an editorial in the *Washington Post* commented on the declining population of migrating birds in the south, attributing this to a bad winter. Carson, believing that the problem was partially caused by pesticide use, sent a response to the newspaper that summarized the arguments made in *Silent Spring*. Her letter ended with the following passage:

> To many of us, this sudden silencing of the song of birds, this obliteration of the color and beauty and interest of bird life, is sufficient cause for sharp regret. To those who have never known such rewarding enjoyment of nature, there should yet remain a nagging and insistent question: If this "rain of death" has produced so disastrous an effect on birds, what of other lives, including our own? (Carson 1959: 26)

On another occasion, she spoke before her neighbors in the Quaint Acres Community Association to give her views on a proposed spraying of the area even though insects were not a major problem. Emphasizing the relationship she detected between cancer and pesticide use, she persuaded the residents to vote against the spraying program. In the summer of 1961, she wrote a letter to the *Boothbay Register* criticizing the insecticide-spraying programs intended to combat Dutch elm disease. She thought her comments would be read by a limited readership, but because of her fame, the letter was transmitted nationally by the wire services.

It was fortunate, in a way, that *Silent Spring* took far longer for Carson to write than she had anticipated because, as Carol Gartner observed, "Just at the right moment, all the elements—writer, subject, audience—came together in synergy to produce a masterpiece" (Gartner 1983: 86). The general public was becoming more aware of problems associated with the use of pesticides, particularly because of the media attention given to "The Great Cranberry Scandal of 1959." Shortly before Thanksgiving 1959, the Food and Drug Administration (FDA) had banned cranberries sprayed with the herbicide aminotriazole before being harvested because the herbicide was only approved for use after harvest. Following studies of the 1958 crop, scientists determined that aminotriazole caused thyroid cancer in rats. This incident "revealed that the laws protecting consumers from toxic chemicals were inadequate and exposed the need for legislation forcing manufacturers to demonstrate the safety of chemicals before, not after, marketing" (Lear 1997: 359–60). Rachel had attended the hearings held in Washington that resulted in the ban on cranberries sprayed with aminotriazole. The media coverage that was directed at this episode not only made the public more aware of chemical misuse but also created an audience that would be more interested in Carson's book.

In 1962, right before the publication of *Silent Spring*, the thalidomide controversy would have a similar effect. It became common knowledge in July 1962 that pharmaceutical companies were considering marketing the drug thalidomide in the United States. Thalidomide, which had been linked to birth defects such as deformed limbs, was used as a sedative and sleeping aid but was dangerous when used by pregnant women. Ultimately, the FDA banned thalidomide because Dr. Frances Oldham Kelsey, an FDA staff member, had been so adamant about its dangers. Carson considered thalidomide and pesticides similar in that they both "represent our willingness to rush ahead and use something new without knowing what the results are going to be" (quoted in Lear 1997: 412). Pesticides were also compared with something that had been causing great public concern for over a decade—radioactive fallout. Just like fallout from nuclear weapons, many scientists believed that pesticides could cause "a kind of pollution that was invisible to the senses; could be transported great distances, perhaps globally; could accumulate over time in body tissues; could produce chronic as well as acute, poisoning; and could result in cancer, birth defects, and genetic mutations that may not become evident until years or decades after exposure" (Lutts 2000: 19).

## CARSON'S HEALTH DETERIORATES

As 1959 drew to a close, Rachel, despite missing her October 1959 deadline, was feeling positive about her work and energized to complete it in the foreseeable future. It was taking much longer than she had originally thought to complete this project, but she wanted to support her writing with sufficient evidence to convince both scientists and laypeople of the validity of her findings. Unfortunately, Rachel began experiencing health problems in January 1960 with an ulcer followed by viral pneumonia and then a sinus infection. In the spring, just as she was feeling well enough to start work on the book once more, she was again diagnosed with breast cancer. The tumor that had been removed in 1950 was incorrectly identified as benign when, in fact, it was malignant. After a radical mastectomy, her doctors misinformed her that no further treatment was necessary, but by the end of the year, the tumor had metastasized, the cancer had spread, and she underwent radiation treatments that weakened her and made writing impossible.

Rachel wrote to Dorothy Freeman, "I suppose as I grow older and become more aware that life is not only uncertain but short at best, the sense of urgency grows to press on with the things I need to say" (quoted in Freeman 1995: 310–11). By the end of the year, she also suffered from a bacterial infection that was diagnosed by one of her doctors as infectious rheumatoid arthritis. Because of the pain in her joints, she was left bedridden or in a wheelchair and was unable to write for months. In another letter to Dorothy, she wrote hopefully, "I sometimes have a feeling...that perhaps this long period away from active work will give me the perspective that was so hard to attain, the ability to see the woods in the midst of the confusing multitude of trees" (357). After being diagnosed with iritis, she wrote, "Such a catalogue of illnesses! If one were superstitious it would be easy to believe in some malevolent influence at work, determined by some means to keep the book from being finished" (390). Later, she confided, "if my time were to be limited, the thing I wanted above all else was to finish this book" (391).

# Chapter 10

# THE OTHER ROAD

## OTHER PROJECTS

Carson, despite her time constraints, worked on a number of smaller projects while she was writing *Silent Spring*. She backed out of doing *The World of Nature* anthology for Simon & Schuster in favor of a compilation titled *Magic of the Sea* for Harper & Brothers and signed a contract in April 1959. She also wrote the preface to the Animal Welfare Institute's pamphlet titled *Humane Biology Projects* that concerned reforming the instruction of high school biology to avoid inhumane experiments with animals. In a passage that reflects her trademark ecological perspective as well as her adherence to Schweitzer's "reverence for life" philosophy, she wrote, "To understand biology is to understand that all life is linked to the earth from which it came; it is to understand that the stream of life, flowing out of the dim past into the uncertain future, is in reality a unified force, though composed of an infinite number and variety of separate lives" (quoted in Lear 1998: 193).

Carson somehow found time to be politically active during this period because she felt very strongly that a Democratic president would support more favorable environmental policies than Dwight Eisenhower had during his two terms in office. She worked on John Kennedy's 1960 election campaign by serving on the Natural Resources Committee of the Democratic Advisory Council and was named to the Women's Committee for New Frontiers, which also included former First Lady Eleanor Roosevelt and Frances Perkins, who had served as secretary of labor during the Roosevelt administration. After Kennedy was elected, Carson attended the

Distinguished Ladies Inaugural Reception but had to decline her invitation to the Inaugural Ball.

Carson also worked on a revision of *The Sea Around Us* to incorporate new discoveries made by oceanographers since the first edition of the book was published in 1951. Beginning in 1960, the record-breaking descents of the bathyscaphe *Trieste* had brought divers to the deepest parts of the ocean in the Mariana Trench almost seven miles below the surface. To bring her book up-to-date, Carson added lengthy footnotes, wrote a new preface providing an overview of the major advancements in oceanography and marine biology, and updated the appendix. In the preface, which reflects a new perspective brought about by the Cold War, Carson expressed grave concerns about the disposal of atomic materials in the ocean:

> Although man's record as a steward of the natural resources of the earth has been a discouraging one, there has long been a certain comfort in the belief that the sea, at least, was inviolate, beyond man's ability to change and to despoil. But this belief, unfortunately, has proved to be naïve. In unlocking the secrets of the atom, modern man has found himself confronted with a frightening problem—what to do with the most dangerous materials that have ever existed in all the earth's history, the by-products of atomic fission. The stark problem that faces him is whether he can dispose of these lethal substances without rendering the earth uninhabitable. (Carson 1961: xi)

On June 12, 1962, Carson delivered a commencement address at Scripps College in Claremont, California, that echoed the fears expressed in her preface to the revised edition of *The Sea Around Us* published the year before. Although her physical condition was deteriorating and the trip would be exhausting, she wanted to take advantage of the opportunity to affirm her views on the relationship of man and nature. In explaining her reason for accepting the invitation to speak, she wrote to Dorothy Freeman that she had developed "a deepened awareness of the preciousness of whatever time is left, be it long or short, and a desire to live more affirmatively, making the most of opportunities when they are offered, not putting them off for another day" (quoted in Freeman 1995: 332).

In her speech, "Of Man and the Stream of Time," Carson shared these reflections:

> I used to wonder whether nature . . . actually needed protection from man. Surely the sea was inviolate and forever beyond

man's power to change it. Surely the vast cycles by which water
is drawn up into the clouds to return again to the earth could
never be touched. And just as surely the vast tides of life—the
migrating birds—would continue to ebb and flow over the
continents, marking the passage of the seasons. But I was
wrong. Even these things, that seemed to belong to the eternal
verities, are not only threatened but have already felt the destroying hand of man. (quoted in Gartner 1983: 120)

In closing she encouraged the young graduates, "Your generation must
come to terms with the environment. Your generation must face realities
instead of taking refuge in ignorance and evasion of truth. Yours is a grave
and a sobering responsibility, but it is also a shining opportunity" (121).

## COMPLETING *SILENT SPRING*

At the beginning of 1962, after Carson had completed fifteen of *Silent
Spring's* seventeen chapters, she sent the manuscript to Paul Brooks and a
copy to William Shawn at the *New Yorker*, who hailed it as a "brilliant
achievement" (quoted in Lear 1997: 395). Brooks told Rachel, "I cannot
imagine anyone else who possesses the combination of scientific understanding and literary skill to make such a fine book out of such difficult
and complicated material" (396). Rachel wrote to Dorothy, "last night
the thoughts of all the birds and other creatures and all the loveliness that
is in nature came to me with such a surge of deep happiness, that now I
*had* done what I could—I had been able to complete it—now it had its
own life!" (quoted in Freeman 1995: 394). By April she had managed to
finish the two incomplete chapters as well as revise the third chapter so it
would not be too technical for the layperson, despite the discovery the
month before that her cancer had spread, requiring her to undergo further
radiation treatments.

In addition to having problems writing her book, Rachel had a difficult
time choosing a title for it. Concerning this decision, she wrote to
Dorothy, "I told you that a possible opening sentence had drifted to the
surface of my mind recently. It was—'This is a book about man's war
against nature, and because man is part of nature it is also and inevitably
a book about man's war against himself'" (quoted in Freeman 1995: 380).
She then mentioned possible book titles: *The War against Nature* and *At
War with Nature*. Other titles that were considered were *How to Balance
Nature, The Control of Nature, Man against the Earth,* and *Dissent in Favor
of Man*. Marie Rodell, who joked that it should just be called *Carson:*

*Opus #4*, suggested that the book be titled *Silent Spring* even though that was originally just the title of one of the chapters. After coming up with no better title than *Man against the Earth*, Paul Brooks began to feel that *Silent Spring* reflected the theme of the entire book and convinced Carson of this. After the book became a classic, it was hard to imagine it being called anything else.

## AN OVERVIEW OF *SILENT SPRING*

Rachel dedicated *Silent Spring* to Albert Schweitzer, who won the Nobel Peace Prize in 1952 and was a strong opponent of nuclear weapons, paraphrasing his words in the dedication, "Man has lost the capacity to foresee and forestall. He will end by destroying the earth." In the front matter of the book, she also included a line from a John Keats poem that provided the inspiration for the title: "The sedge is wither'd from the lake, / And no birds sing." Finally she quoted E. B. White because his views so closely resembled her own: "I am pessimistic about the human race because it is too ingenious for its own good. Our approach to nature is to beat it into submission. We would stand a better chance of survival if we accommodated ourselves to this planet and viewed it appreciatively instead of skeptically and dictatorially." A lengthy list of acknowledgments, which begins by crediting Olga Owens Huckins's letter with motivating Carson to write the book, expresses gratitude to the many experts who reviewed chapters from her manuscript and lends added credibility to her work.

The opening chapter of *Silent Spring*, "A Fable for Tomorrow," describes a town that bears a striking resemblance to Carson's hometown of Springdale, Pennsylvania, during her childhood: "There was once a town in the heart of America where all life seemed to live in harmony with its surroundings. The town lay in the midst of a checkerboard of prosperous farms, with fields of grain and hillsides of orchards where, in spring, white clouds of bloom drifted above the green fields" (Carson 1962: 1). Following a description of the beauty of the other seasons and the abundance of wildlife, it is observed that "a strange blight crept over the area and everything began to change" (2). Illness and death came upon all the living creatures including humans, and the vegetation withered away. A "white granular powder...had fallen like snow" on the community a few weeks before, obviously a pesticide (3). While admitting that this town does not truly exist, Carson asserted that each of the things that happened there had happened somewhere. Many scientists objected to this allegorical opening, but because Carson "knew that her book must persuade as well as inform; it must synthesize scientific fact with the most profound sort of propaganda," she felt that her "fable" would draw the general reader more deeply into the book (Graham 1970: 22).

Carson then provided an overview of man's abuse of the environment in a chapter titled "The Obligation to Endure." She emphasized the delicate balance of nature that has taken millions and millions of years to develop and contrasted this with the rapid poisoning of the environment brought about by human beings in the twentieth century due, in part, to the use of radiation and DDT. To clarify her position that the banning of all pesticides is not necessary, she wrote, "It is not my contention that chemical insecticides must never be used. I do contend that we have put poisonous and biologically potent chemicals indiscriminately into the hands of persons largely or wholly ignorant of their potentials for harm" (Carson 1962: 12). She particularly objected to the indiscriminate use of pesticides, which she referred to as "biocides," without the knowledge or consent of those affected by the chemicals (8).

Chapter 3, "Elixirs of Death," was a difficult one for Carson to write because she was concerned that general readers would become so bogged down in the technical information contained in it that they would fail to appreciate the rest of the book. She felt, however, that a background in the history and description of pesticides as well as herbicides, which are used to destroy plants, was essential to understanding *Silent Spring*. She expressed her dilemma in a letter to Dorothy Freeman, writing, "How to reveal enough to give understanding of the most serious side effects of the chemicals without being technical, how to simplify without error—these have been problems of rather monumental proportions" (quoted in Freeman 1995: 387). Paul Brooks felt that "if the reader could be piloted through chapter 3, the remainder would be comparatively smooth sailing" (Brooks 1972: 267). This chapter introduces the reader to herbicides and the two main groups of pesticides: the chlorinated hydrocarbons, which break down slowly and include DDT, chlordane, dieldrin, aldrin, endrin, and helptachlor, and the organic phosphates, which are highly toxic but do not remain in the environment for a long time, including parathion and malathion.

Beginning with chapter 4, "Surface Waters and Underground Seas," which is concerned with water pollution, a series of chapters describe the effects of pesticides. Carson asserted, "It is not possible to add pesticides to water anywhere without threatening the purity of water everywhere" (Carson 1962: 42). She explained that the contamination of this most basic resource threatens the entire food chain: "Water must also be thought of in terms of the chains of life it supports—from the small-asdust green cells of the drifting plant plankton, through the minute water fleas to the fishes that strain plankton from the water and are in turn eaten by other fishes or by birds, mink, raccoons—in an endless cyclic transfer of materials from life to life" (46). A case study of Clear Lake, California, demonstrates this point. Plankton in Clear Lake that had absorbed the water were found to contain five parts per million of an insecticide related to DDT known as DDD. Plant-eating fish, however, had as high as three hundred parts per million in their systems, and larger carnivorous fish had levels up to twenty-five hundred parts per million, thus showing that poisons become more concentrated as they are absorbed by creatures higher up the food chain. This concept, known as bioconcentration, was important to Carson's argument about the dangers of pesticides because human beings occupy a terminal point in the food chain.

After a chapter that analyzes the contamination of another basic resource—soil—the effects of herbicides, which are related to pesticides in that they kill living organisms, are examined in "Earth's Green Mantle." Rachel criticized the use of weed killers to destroy plants that do not seem to serve a purpose, explaining that no one can know how a certain plant might fit the "web of life in which there are intimate and essential relations between plants and the earth, between plants and other plants, between plants and animals" (Carson 1962: 64). She lamented the "brown desolation" of the roadsides of Maine leading to her Southport Island cottage that had once been lined with an abundance of plant life (70). While civic leaders argued that herbicides were cheaper than mowing the grass and cutting down plants that obstruct the view of drivers, Carson believed there was a greater indirect cost. Not only did the animals that ate the sprayed plants suffer ill effects, but the beauty of the landscape was destroyed.

Chapter 7, "Needless Havoc," describes "the direct killing of birds, mammals, fishes, and indeed, practically every form of wildlife by chemical insecticides indiscriminately sprayed on the land" and provides two case studies to illustrate this tragedy (Carson 1962: 85). In the fall of 1959, the Detroit area was sprayed with aldrin, the deadliest and cheapest pesticide, to eradicate Japanese beetles, even though these insects were not causing a serious problem. After the spraying, the local health department received a steady stream of calls from people complaining of coughing, nausea, vomiting, fever, chills, and fatigue. Many birds were dead, and a large percentage of the cat and dog population was ill. Attempts were also made to eliminate this same insect species in Sheldon, Illinois, from 1954 through 1961 by the U.S. Department of Agriculture and the Illinois Agriculture Department, resulting in "unparalleled wildlife destruction" (92). Instead of permanently eradicating the beetles, some species of insect-eating birds were destroyed. Ninety percent of cats died during the first season of spraying, and livestock including sheep and cattle were also killed. Carson concluded the chapter thus: "The question is whether any civilization can wage relentless war on life without destroying itself, and without losing the right to be called civilized" (99).

Chapter 8, "And No Birds Sing," opens with this haunting passage that inspired the title of the book:

> Over increasingly large areas of the United States, spring now comes unheralded by the return of the birds, and the early mornings are strangely silent where once they were filled with the beauty of bird song. This sudden silencing of the song of birds, this obliteration of the color and beauty and interest they lend to our world have come about swiftly, insidiously, and unnoticed by those whose communities are as yet unaffected. (Carson 1962: 103)

This chapter describes the massive spraying of DDT throughout the United States beginning in 1954 to control elm bark beetles. These insects were causing Dutch elm disease, threatening the existence of the stately American elms that had been planted in great numbers along the streets and in the parks of many towns. Instead of killing the beetles, however, the DDT killed robins and many other species of birds that died after eating earthworms contaminated with the pesticide. Carson explained that birds are the natural predators of insects, and although insect populations may be reduced temporarily by pesticides, when their numbers increase again, there may be few birds left to control insect population growth. Although the focus of this chapter is on the robin's fate, the eagle is also discussed. Eagles were threatened with extinction because their ability to reproduce had been negatively impacted by DDT. As the chapter closes, this rhetorical question is asked: "Who has decided...that the supreme value is a world without insects, even though it be also a sterile world ungraced by the curving wing of a bird in flight?" (127).

After her discussion of the effects of pesticides on birds, Carson examined the negative impact of these substances on fish in "Rivers of Death." She cited numerous examples of the fish population being devastated by pesticides, including the infamous "fire ant program" in the southern United States and situations that had occurred in such diverse places as Maine, Yellowstone National Park, and British Columbia. Carson described the attempts to eradicate the fire ant in more detail in the following chapter, "Indiscriminately from the Skies," which also discusses aerial spraying to control the gypsy moth in the northeastern United States. She criticized "the end-justifies-the-means philosophy" of the USDA and called the program to control fire ants "an outstanding example of an ill-conceived, badly executed, and thoroughly detrimental experiment in the mass control of insects" (Carson 1962: 156, 162).

The cryptic title of chapter 11, "Beyond the Dreams of the Borgias," is an allusion to the Borgia family—a political dynasty in fifteenth- and sixteenth-century Italy who were notorious for poisoning their enemies. Carson revealed that human beings living in what she calls "the age of poisons" are not only threatened by pesticides sprayed from the sky or contaminating the water, but by the substances available to every citizen for use in the kitchen and the garden (Carson 1962: 174). Home owners who buy pesticides and herbicides that do not have adequate warnings in an effort to get rid of insects in their houses and crabgrass in their lawns "are in little better position than the guests of the Borgias" (184).

Chapters 12 through 14, considered the most controversial part of the book, are concerned with the health effects of pesticides on human beings. "The Human Price" specifically examines "the ecology of the world within our bodies" (Carson 1962: 189) and the long-term effects of pesticides stored in fatty tissues, as well as in the human liver and nervous system. "Through a Narrow Window" examines the changes that pesticides cause on the cellular level, including decreased energy levels, and even speculates that pesticides may result in birth defects. The chapter "One in Four," which Carson originally thought would be part of a chapter on the general physical effects of pesticides on humans, links pesticides with cancer. As she began to appreciate the importance of this subject, she realized she would have to devote an entire chapter to it.

Chapters 15 through 17 suggest some solutions to the problems caused by pesticides. These chapters also explain how insects develop resistance to pesticides and how humans have upset the "balance of nature," which Carson defined as "a complex, precise, and highly integrated system of relationships between living things" (Carson 1962: 246). Although some critics contended that DDT was responsible for nearly wiping out malaria, Carson felt that this was a short-lived victory because mosquitoes would eventually develop resistance to the pesticides and return in even larger numbers. Although a chemical kills the vast majority of insects that are its intended target, a few of the pests remain unaffected. These insects, which possess the gene for resistance, survive to reproduce. Their offspring are more likely to have the same gene for resistance, and these offspring in turn reproduce so that over time, resistance in the entire population is strengthened. The development of resistance is an example of the way in which the process of natural selection works.

Throughout *Silent Spring*, Carson recommended "biological controls" that mimic the types of controls applied by nature. For example, she suggested removing the elm wood where beetles breed to prevent Dutch elm disease and growing a healthy lawn to prevent crabgrass. Other biological methods she recommended include predatory insects and parasites, insect sterilization, and ultrasonic sound. She reiterated her belief that the use of chemical pesticides does not have to be stopped completely, just the use of chlorinated hydrocarbons such as DDT.

Carson began the final chapter of *Silent Spring*, "The Other Road," with these words:

> We stand now where two roads diverge. But unlike the roads in Robert Frost's familiar poem, they are not equally fair. The road we have long been traveling is deceptively easy, a smooth superhighway on which we progress with great speed, but at its end lies disaster. The other fork of the road—the one "less traveled by"—offers our last, our only chance to reach a destination that assures the preservation of our earth. (Carson 1962: 277)

This allusion to Robert Frost's poem illustrates perfectly her belief that the easiest way to eradicate pests and weeds is not the best way and that to survive society needs to reconsider its reliance on pesticides as well as herbicides.

Having decided against including footnotes throughout the text of *Silent Spring* because this format would probably be off-putting to the general reader, Carson provided fifty-five pages of documentation at the end of the book to support her controversial claims. She later said that "in *Silent Spring* I have never asked the reader to take *my* word. I have given him a very clear indication of my sources.... This is the reason for the 55 pages of references. You cannot do this if you are trying to conceal or distort or to present half truths" (quoted in Lear 1998: 207). Carson's voluminous endnotes reflect her years of research.

## THE SIGNIFICANCE OF *SILENT SPRING*

*Silent Spring* was a great departure from Carson's earlier works, not only because it was not about the ocean, as *Under the Sea-Wind*, *The Sea Around Us*, and *The Edge of the Sea* had been, but because it contained such controversial material about such an urgent topic. Cornelius Browne noted, "In her earlier writing, Carson celebrated ideas of ecological interdependence, interrelationships, food chains, and webs. But now she reveals a dark side of interrelationship: the cycles of life that she so celebrated now undeniably contain billions of tons of man-made poisons" (Browne 2002: 33). At the same time, Silent Spring "would not have been possible without the previous books. It is the sense of wonder about nature and the web of life that ties all the books together" (McCay 1993: 108).

Critics agreed that Carson did not present any original scientific research. She acknowledged this when she wrote to Dorothy Freeman about her work in general: "I consider my contributions to scientific fact far less important than my attempts to awaken an emotional response to the world of nature" (quoted in Freeman 1995: 231). What made Silent Spring unique was the way in which Carson gathered together the evidence that supported the case against pesticides and presented this case in a convincing way. Carol Gartner asserted, "Her ability to present complex scientific material with beauty as well as clarity greatly contributes to the effectiveness of Silent Spring" (Gartner 1983: 91–92). In addition, the basic science explained in Silent Spring has not been superseded by more recent discoveries; instead, these discoveries seem to lend more support to Carson's findings.

Patricia Hynes referred to Silent Spring as "the most vital and controversial book ever written on the environment" and further asserted that it "altered a balance of power in the world. No one since would be able to sell pollution as the necessary underside of progress so easily or uncritically" (Hynes 1989: 2, 3). Silent Spring brought terms such as "ecology," "interdependence," and "balance of nature" into common usage. Norman Boucher asserted, "The notion that chemicals can act like radiation; that poisons can climb the food chain, becoming more concentrated as they ascend; that chemicals can even upset chromosomes; that pollution can linger long after its source has been eliminated; that chemicals can seriously contaminate ground water—these ideas, amounting almost to a summary of today's scientific assumptions concerning pollution, are just a few of those first clearly suggested and linked in Silent Spring" (Boucher 1987: 38). Whereas every other book published about chemical pesticides before Silent Spring focused on the economic rather than ecological impact that they had had and mostly advocated their unrestricted use to aid in agriculture, Carson focused on the environmental ramifications.

Carson's collection of case studies took years of painstaking research. While she was struggling to complete Silent Spring, she wrote to Paul Brooks, "I guess all that sustains me is a serene inner conviction that when, at last, the book is done, it is going to be built on an unshakable foundation. That is so terribly important. Too many people—with the best possible motives—have rushed out statements without adequate support, furnishing the best possible targets for the opposition" (quoted in Brooks 1972: 258). Carson's thoroughness would serve her well during the firestorm that followed the publication of *Silent Spring*, when she was forced to defend her claims before the chemical industry, government officials, the press, and the world.

# Chapter 11

# THE LOUD REACTION
# TO *SILENT SPRING*

## ANOTHER BESTSELLER PREVIEWED
## IN THE *NEW YORKER*

A headline in the *New York Times* on July 22, 1962, declared "*Silent Spring* is Now Noisy Summer," reflecting the intense reaction to the *New Yorker*'s condensed version of Carson's book. The magazine, which published the first of three installments on June 16, 1962, received more mail about the *Silent Spring* excerpts than it had ever gotten for any other article. Most of the letters were positive, and many thanked Carson for exposing such a serious problem. A few letters, however, gave a foretaste of the criticism she would experience after the book's publication. One reader accused her of having "Communist sympathies" and even wrote, "isn't it just like a woman to be scared to death of a few little bugs!" (quoted in Glotfelty 2000: 157–58).

The appearance of *Silent Spring* in the *New Yorker* brought attention to the problem of pesticides on a national level. Congressman John V. Lindsay of New York read part of the last installment to the House of Representatives when discussing the issue of pest control. At an August 29, 1962, White House press conference, a reporter asked President John F. Kennedy, "Mr. President, there appears to be a growing concern among scientists as to the possibility of dangerous long-range side effects from the use of DDT and other pesticides. Have you considered asking the Department of Agriculture or the Public Health Service to take a closer look at this?" Kennedy replied, "Yes, and I know they already are. I think particularly, of course, since Miss Carson's book, but they are examining the matter" (quoted in Brooks 1972: 305). By the end of 1962, due in large part to Carson's warnings, more than forty bills had been introduced in state legislatures related to pesticide usage regulation. *Silent Spring* was influential outside the United States as well and was translated into numerous languages.

When Carson first began writing *Silent Spring*, Clarence Cottam believed that the book would "render a great public service" (quoted in Lear 1997: 336). He erroneously predicted, however, that it would not be a bestseller as her other books had been. Following the publication of *Silent Spring* on September 27, 1962, Carson proved him quite wrong about the popularity of her book, which held first place on the *New York Times* bestseller list for most of the fall of 1962 and remained on the list for a total of thirty-one weeks. It was the Book-of-the-Month Club selection for October, and by December 1962 more than one hundred thousand copies had been sold. Half a million hardcover copies were sold before a paperback edition was released. The public paid attention to what Carson had to say and bought *Silent Spring* in great numbers because she had established herself as an authority through her other books. Carol Glotfelty observed, "When Carson's tune suddenly changed from one of gentle appreciation to one of grim alarm, people listened" (Glotfelty 1996: 161).

## THE OPPOSITION TO *SILENT SPRING*

Although *Silent Spring* met a receptive public, it was, not surprisingly, denigrated by the chemical industry that produced the extremely profitable pesticides. Paul Brooks compared Carson's book to Charles Darwin's treatise on evolution, *The Origin of Species*, and acknowledged, "Perhaps not since the classic controversy over Charles Darwin's *The Origin of Species* just over a century earlier had a single book been more bitterly attacked by those who felt their interests threatened" (Brooks 1972: 293). The Velsicol Chemical Corporation, the only producer of the chlorinated hydrocarbons chlordane and heptachlor, objected to the book's publication because of alleged inaccuracies concerning the company's products and tried to persuade Houghton Mifflin not to release it by threatening a libel suit. After the serialization in the *New Yorker*, the company sent a letter to the book's publisher that revealed the influence of Cold War politics. It went so far as to suggest that *Silent Spring* was part of an evil Communist plot to destroy the agriculture and economy of the country, stating that "members of the chemical industry in this country and in Western Europe must deal with sinister influences, whose attacks on the chemical industry have a dual purpose: (1) to create the false impression that all business is grasping and immoral, and (2) to reduce the use of agricultural chemicals in this country and in the countries of western Europe, so that our supply of food will be reduced to east-curtain parity" (quoted in Graham 1970: 49).

The Monsanto Corporation wrote a parody of Carson's opening chapter, "A Fable for Tomorrow," titling it "The Desolate Year." This account of a world without pesticides in which insects cause death and destruction was sent to newspapers nationwide. The National Agricultural Chemicals Association, the trade organization for the pesticide industry, spent more than a quarter million dollars to conduct a public relations campaign against Carson. Paul Brooks credited the attacks by the pesticide industry with providing "infinitely more publicity than Houghton Mifflin could ever have afforded" (quoted in "Rachel Carson's *Silent Spring*" 1993). Those in the chemical industry who criticized the book often tried to undermine Carson's credibility by focusing on her gender and her alleged lack of professionalism because she wrote for a general audience and did not possess a doctorate in science. Paul Brooks suggested, "Her opponents must have realized—as was indeed the case—that she was questioning not only the indiscriminate use of poisons but the basic irresponsibility of an industrialized, technological society toward the natural world" (Brooks 1972: 293).

Criticism of Carson was not restricted to the chemical industry. Numerous government officials disapproved of *Silent Spring*. Ezra Taft Benson, former secretary of agriculture, accused Carson of being a Communist. Congressman Jamie L. Whitten of Mississippi, chair of the House Appropriations Subcommittee on Agriculture, disagreed heartily with Carson's views. He promoted the use of pesticides in his 1966 book *That We May Live*, in which he wrote:

> I believe all must agree that *Silent Spring*, delightful reading that it is, certainly is not and was never claimed to be a scientific document nor an objective analysis of the chemical-human life relationship. Though we give it our highest praise for its wonderful prose, for its timely warning, let us move it over from the non-science fiction section of the library to the science-fiction section, while we review the facts—in order that we may continue to enjoy the abundant life. (Whitten 1966: 141)

One member of the Federal Pest Control Board described Carson as a "spinster" and asked, "What's she so worried about genetics for?" (quoted in Graham 1970: 50). According to F. A. Soraci, the director of the New Jersey Department of Agriculture, Carson belonged to "a vociferous, misinformed group of nature-balancing, organic-gardening, bird-loving, unreasonable citizenry that has not been convinced of the important place of agricultural chemicals in our economy" (56). Some of her critics hadn't even read the book. A newspaper article in the *Globe-Times* of Bethlehem, Pennsylvania, reported, "No one in either county farm office who was talked to today had read the book, but all disapproved of it heartily" (48).

One unlikely adversary was Dr. Cynthia Westcott, an entomologist and writer of a popular gardening column in *American Woman*, "The Plant Doctor," who had also been the chairperson of the National Council of State Garden Clubs. In her review of *Silent Spring* for the *National Gardener*, she questioned the health threats posed by pesticides. Considering Carson to be an alarmist who manipulated facts to serve her purposes, Westcott continued to advocate the benefits of pesticides to the gardener.

Unlike the universal critical acclaim given her previous books, *Silent Spring* was met with a negative reaction by numerous magazine reviewers. Some companies threatened to pull their ads from publications that gave good reviews to *Silent Spring*. *Reader's Digest, Science,* and *Time* initially published unfavorable reviews of *Silent Spring*. *Reader's Digest* cancelled a contract to publish a twenty-thousand-word condensation. A review in *Time* magazine stated, "Miss Carson has taken up her pen in alarm and anger, putting literary skill second to the task of frightening and arousing readers" (*Time* 1962: 45). It also asserted that Carson's case was "unfair, one-sided, and hysterically overemphatic" and that her book contained "oversimplifications and downright errors" (48). The reviewer reported that the evidence had shown DDT to be relatively harmless to humans if used properly and that damage to wildlife "though always regrettable, is not disastrous." Finally, he wrote, "Many scientists sympathize with Miss Carson's love of wildlife, and even with her mystical attachment to the balance of nature. But they fear that her emotional and inaccurate outburst in *Silent Spring* may do harm by alarming the nontechnical public, while doing no good for the things that she loves."

An article in the *Economist* argued that Carson's "angry, shrill tract" was "propaganda written in white-hot anger with words tumbling and stumbling all over the page" (*Economist* 1963: 711). George Decker's review in *Chemical World News* belittled the book, comparing it to television's *The Twilight Zone*. The *Chemical and Engineering News* stated, "in view of her scientific qualifications in contrast to those of our distinguished scientific leaders and statesmen, this book should be ignored" (Darby 1962: 60).

In September 1963, the *Saturday Evening Post* published "The Myth of the 'Pesticide Menace'" by Edwin Diamond, the senior editor of *Newsweek* who had been dismissed as Carson's coauthor. In this article, Diamond criticized Carson as being an emotional alarmist who employed some of the same tactics as Communist hunter Joseph McCarthy. Carson, along with Paul Brooks and Marie Rodell, felt that Diamond was motivated by revenge to write the hostile and negative review of *Silent Spring*. He was referred to in the magazine as "Miss Carson's collaborator at the start of the project that became *Silent Spring*." Paul Brooks wrote a letter to the editor clarifying that Diamond never actually worked with Carson on the project at all.

## SUPPORT FOR *SILENT SPRING*

A number of critics responded to *Silent Spring* more positively. Although a mixed review by Lorus and Margery Milne that appeared on the front page of the *New York Times Book Review* asserted that *Silent Spring* was "so one-sided that it encourages argument," it also conceded that "little can be done to refute Miss Carson's carefully documented statements" (Milne 1962: 1). Robert Cowen shared this opinion in the *Christian Science Monitor*, writing, "Miss Carson has undeniably sketched a one-sided picture. But her distortion is akin to that of the painter who exaggerates to focus attention on essentials. It is not the half-truth of the propagandist" (Cowen 1962: 11). A review in the *Christian Century* asserted, "Miss Carson's scientific credentials are impeccable" and that *Silent Spring* "is a shocking and frightening book. It ought to be placed on the required reading list of every community leader, every lover of nature, and every citizen who cherishes the great natural resources of our nation" (*Christian Century* 1962: 1564). Loren Eiseley of the *Saturday Review* stated, "It is a devastating, heavily documented, relentless attack upon human carelessness, greed, and irresponsibility" and added, "If her present book does not possess the beauty of *The Sea Around Us*, it is because she has courageously chosen, at the height of her powers, to educate us upon a sad, an unpleasant, an unbeautiful topic, and one of our own making" (Eiseley 1962: 18). Marston Bates of the *Nation* praised Carson for making "a real contribution to our salvation" (Bates 1962: 202).

Many authorities compared *Silent Spring* to *Uncle Tom's Cabin*, Harriet Beecher Stowe's popular novel published in 1852, which contributed to the abolition of slavery in the United States. Supreme Court Justice William O. Douglas called *Silent Spring* "the most revolutionary book since *Uncle Tom's Cabin*" as well as "the most important chronicle of this century for the human race" (quoted in Lear 1997: 419; Bonta 1991: 262). E. B. White reiterated this sentiment when he wrote to Rachel, "This will be an *Uncle Tom's Cabin* of a book,—the sort that will help turn the tide" (quoted in Lear 1997: 421). During Senate hearings on the use of pesticides, Senator Ernest Gruening also compared the potential effect of *Silent Spring* to that of *Uncle Tom's Cabin* and thought that it, too, could change history. Author Robert B. Downs wrote about *Silent Spring* as well as *Uncle Tom's Cabin* in *Books That Changed America*, published in 1970. Vice President Al Gore, in his introduction to the 1994 edition of *Silent Spring*, pointed out that there was an important difference between Stowe's book and Carson's. Slavery was already a hotly debated issue in the 1850s while the danger of pesticides was a problem that hardly anyone knew about in the 1960s. Gore observed that Carson "was trying to put an issue on the national agenda, not bear witness to one that was already there" (Gore 1994: xix).

## AWARDS FOR *SILENT SPRING*

Despite the negative reaction from industry and critics, Carson won numerous awards following the publication of *Silent Spring*. Her most coveted award was the Albert Schweitzer Medal of the Animal Welfare Institute, which she received in 1963. In her acceptance speech, she said, "I can think of no award that would have more meaning for me or that would touch me more deeply than this one, coupled as it is with the name of Albert Schweitzer." Acknowledging the importance of his "reverence for life" philosophy, she asserted, "If, during the coming years, we are to find our way through the problems that beset us, it will surely be in large part through a wider understanding and application of his principles" (quoted in Brooks 1972: 315). Carson also received the Conservationist of the Year award from the National Wildlife Federation, the Conservation Award of the Izaac Walton League of America, and the Cullum Medal from the American Geographical Society. She was the first woman to win the Audubon Medal from the National Audubon Society and also received the National Council of Women's first Woman of Conscience award. Compared to Galileo by the American Academy of Arts and Letters, she was elected to this organization of fifty members, which included only three other women, and considered her membership "the most deeply satisfying thing that has ever happened in the honors department" (quoted in Lear 1997: 472).

Rachel found it difficult to appreciate the success of her book, however, because of her ongoing health problems, including her battle with cancer.

*Silent Spring* was "the book that gave her the most fame and the least plea-
sure" (Bonta 1991: 270). Right before Christmas in 1962, she had to
begin another round of radiation treatments that lasted several weeks and
had to endure this process again in February when it became apparent
that the cancer had entered her bones. Having also developed angina, she
was advised not to give any more speeches until this condition was under
control. She accepted the fact that her days were numbered, writing to
Dorothy, "I have had a rich life, full of rewards and satisfactions that come
to few, and if it must end now, I can feel that I have achieved most of what
I wished to do" (quoted in Freeman 1995: 542).

Despite her ill health, in January 1963 Carson addressed the Garden
Club of America from which she had received the Frances K. Hutchinson
Medal ten years earlier. She exhorted her audience:

> If we are ever to find our way out of the deplorable situation,
> we must remain vigilant, we must continue to challenge and to
> question, we must insist that the burden of proof is on those
> who would use these chemicals to prove the procedures are
> safe. Above all, we must not be deceived by the enormous
> stream of propaganda that is issuing from the pesticide manu-
> facturers and from industry-related—although ostensibly inde-
> pendent—organizations. (quoted in Lear 1998: 218)

## THE VINDICATION OF *SILENT SPRING*

Rachel Carson testified before the President's Science Advisory Committee (PSAC) in January 1963. The PSAC's report on the effects of pesticides was released on May 15, 1963, and was titled simply "Use of
Pesticides." This report, in addressing both the benefits and hazards of
chemical pesticides, gave a perspective not found in previous government
reports, which had always supported pesticide use. Although it recommended continued use of some toxic chemicals, "Use of Pesticides" supported Carson's findings that DDT and other pesticides had infiltrated the
tissues of humans, wildlife, and the environment on a large scale, causing
untold damage. It also criticized the fire ant program and other eradication
attempts and supported the use of biological controls such as those described in *Silent Spring*. The report, which concluded, "Elimination of the
use of persistent toxic pesticides should be the goal," included a statement
by President Kennedy indicating that he would consider new legislation to
support the report's recommendations (quoted in Graham 1970: 78).

In the PSAC report, Carson was credited with enlightening society about the dangers of pesticides. It acknowledged that "until the publication of *Silent Spring* by Rachel Carson, people were generally unaware of the toxicity of pesticides" (quoted in Graham 1970: 78–79). "Use of Pesticides" was regarded by many journalists and scientists as a vindication of Rachel Carson and also as a reflection of how strongly she had influenced government policy.

On April 3, 1963, *CBS Reports* aired "The Silent Spring of Rachel Carson," an hour-long documentary hosted by Eric Sevareid and Jay McMullen. At the last minute, three of the show's five sponsors cancelled their ads. Lehn & Fink Products, the makers of Lysol, and two food manufacturers, Standard Brands and Ralston Purina, thought that the program would not be an appropriate one on which to promote their products. In the program, Carson, who "appeared anything but the hysterical alarmist that her critics contended," was opposed primarily by the "wild-eyed, loud-voiced Dr. Robert White-Stevens" of the American Cyanamid Company, who seemed much more frenetic than the composed Carson (Lear 1997: 449).

White-Stevens asserted, "the major claims of Miss Rachel Carson's book *Silent Spring* are gross distortions of the actual facts, completely unsupported by scientific experimental evidence and general practical experience in the field," but he was not able to give even one example of a specific inaccurate statement in the book. He argued, "Miss Carson maintains that the balance of nature is a major force in the survival of man whereas . . . the modern scientist believes that man is steadily controlling nature." Carson replied, "Now to these people apparently the balance of nature was something that was repealed as soon as man came on the scene. Well you might just as well assume that you could repeal the law of gravity."

Also appearing on the program were Orville Freeman, the secretary of agriculture; Dr. Luther Terry, the surgeon general; and George Larrick, commissioner of the Food and Drug Administration. Although at first they defended the use of pesticides, by the end of the program they were acknowledging the dangers of these substances. When asked if the public was adequately informed about the hazards of pesticides, Secretary Freeman stated, "The answer I can say very quickly is 'no.'" Commissioner Larrick conceded, "I don't think, in all honesty, that the controls that we're able to exercise today are truly sufficient" and expressed particular concern over widespread aerial spraying. The surgeon general was concerned about "what happens with low-level, long-range exposure of human beings" to pesticides. Sevareid noted that there was "acknowledgment on all sides that a pesticide problem does exist," and McMullen lamented that "scientist after scientist has pointed to an appalling scarcity of facts concerning the effects of pesticides on man and his environment."

Carson, who appeared to be a knowledgeable and concerned scientist rather than the alarmist that critics had called her, said in her characteristic calm tone, "It is not my contention that chemical pesticides must never be used. I do contend that we have allowed these chemicals to be used with little or no advance investigation of their effect on soil, water, wildlife, and man himself." She asserted, "Man is a part of nature, and his war against nature is inevitably a war against himself." At the end of the program she said, "Now I truly believe that we in this generation must come to terms with nature, and I think we're challenged, as mankind has never been challenged before, to prove our maturity and our mastery, not of nature but of ourselves."

The show, which was watched by an audience of more than 10 million people, was a triumph for Carson, who "came across as a dignified, polite, concerned scientist, with no motive other than to alert the public to a significant problem" (Lear 1997: 449). After the PSAC report and the CBS documentary, many magazines, including *Time*, that had initially given negative reviews to *Silent Spring* reconsidered their positions and published new, more positive reviews.

The day after *CBS Reports*, Senator Abraham Ribicoff of Connecticut announced that Congress would hold hearings on pesticide use. He was appointed to chair a subcommittee to review all federal environmental policies pertaining to pesticides and other pollutants. He later asserted, "If it wasn't for Rachel Carson, I never would have had these hearings. I was not aware of the extent and the importance of the problem she raised" (quoted in "Rachel Carson's *Silent Spring*" 1993). "The Ribicoff Report" resulting from the hearings had the same effect as the PSAC report and the CBS documentary in vindicating Carson.

When Carson testified before the Senate subcommittee on June 4, 1963, Ribicoff greeted her by saying, "Miss Carson, you are the lady who started all this," echoing the words of President Abraham Lincoln, who, when meeting Harriet Beecher Stowe a century before, said, "So you're the little lady who started this whole thing" (quoted in Gore 1994: xix). During her forty-minute testimony, she recommended that aerial pesticide spraying be minimized and that the most harmful pesticides should be eliminated altogether. As a result of her government experience, she knew that she had to remain realistic about the types of reforms that the Senate would consider, so she avoided anything controversial in the interest of having the lawmakers take her more seriously. She called for the right of citizens to protection against environmental poisons in their own homes, restrictions on the sale and use of pesticides, and support of research into alternatives to chemical pesticides. She advised, "Since our problems of pest control are numerous and varied, we must search, not for one superweapon that will solve all our problems, but for a great diversity of armaments, each precisely adjusted to its task" (quoted in Graham 1970: 87). Linda Lear wrote, "Those who heard Rachel Carson that morning did not see a reserved or reticent woman in the witness chair but an accomplished scientist, an expert on chemical pesticides, a brilliant writer, and a woman of conscience who made the most of an opportunity few citizens of any rank can have to make their opinions known. Her witness had been equal to her vision" (Lear 1997: 454).

Just two days later, Carson testified before the Senate Committee on Commerce, which was considering the Chemical Pesticides Coordination Act. This legislation mandated stronger pesticide warnings and would also require the USDA to consult the Department of the Interior as well as individual state governments before implementing a pesticide program. Carson recommended a cabinet-level agency that would regulate pesticides and other environmental hazards and, as an independent entity, would not be influenced by the chemical industry. After completing her Senate testimony, Rachel went to Maine for what would be her last summer. Ever mindful that the end was near for her, she wrote to Dorothy, "There is still so much I want to *do*, and it is hard to accept that in all probability, I must leave most of it undone" (quoted in Freeman 1995: 490).

## "THE CLOSING JOURNEY"

Although Rachel Carson's physical condition continued to deteriorate, she remained occupied with several writing projects. In addition to a lengthy essay that she wrote in 1963 for the *World Book Encyclopedia Yearbook* about the resources of the ocean, she also contributed the foreword to Ruth Harrison's *Animal Machines* published the following year, a book about the inhumane treatment of livestock. As someone with an ecological viewpoint who was concerned about the relationship of living things to their environment, she, like Harrison, found the conditions under which animals were raised to be deplorable. She blamed such "monstrous evils" on a modern society that "worships the gods of speed and quantity, and of the quick and easy profit" (quoted in Lear 1998: 194). Meanwhile, Rachel struggled with *Origin of Life*, which at one time appeared to have the potential to be her most significant work. But her heart was never really in it after she began writing *Silent Spring*. Although she told Ruth Anshen in 1963 that she was giving a number of lectures pertaining to the subject of evolution and wanted them to serve a dual purpose by using them as a foundation for *Origin of Life*, she never did finish the project.

The September/October 1963 issue of *Audubon Magazine* included Rachel's article, "Rachel Carson Answers Her Critics," in which she defended herself as being a "tough-minded realist" rather than a "heedless sentimentalist" as some of her critics were portraying her (Carson 1963: 262). She pointed out some incidents demonstrating that the problems discussed in her book continued to occur and wrote, "the problem of pesticides is not merely the dream of an avaricious author, out to pile up royalties by frightening the public—it is very much with us, here and now" (264). After clarifying that her recommendation was not the elimination of pesticides but moderation in their use, she warned readers not to be deceived by propaganda concerning this controversial issue.

In October 1963, Rachel travelled to San Francisco with Marie Rodell to participate in the Kaiser Foundation Symposium, "Man against Himself." It was a difficult trip because radiation treatments had weakened her, and she was in pain most of the time. But she felt it was an important opportunity to address a group of physicians and explained her use of a cane by claiming that she had arthritis. She remained seated while she delivered her hour-long speech "The Pollution of the Environment" to an enthusiastic audience of fifteen hundred people. It was a notable address not only because it was her last, but because in it she first publicly identified herself as an "ecologist," defining ecology as the science which studies the interrelationship of organisms and their environments. In addition to discussing the pesticide issue, she also addressed the growing problem of radioactive waste disposal in the sea. While in California, Rachel visited Muir Woods and, although she had to be transported in a wheelchair, she fulfilled a long-time dream to see the giant redwoods.

When Rachel returned from her trip to California, sensing that the end of her life was near, she spent time sorting through her personal papers and manuscripts and decided to bequeath this material to Yale University's Beinecke Rare Book and Manuscript Library. On April 14, 1964, at the age of fifty-six, after suffering through what she had called a "catalogue of illnesses," Rachel Carson died of a heart attack (quoted in Freeman 1995: 390). Most of her friends and acquaintances were surprised to learn of her death because she had gone to great lengths to disguise her condition, and they had not known she was so ill. In a codicil to her will, she had named Paul Brooks and his wife or Dorothy Freeman's son and daughter-in-law as her choices to be Roger's guardians. Having never informed her nominees, they were all quite surprised, but Paul Brooks agreed to take care of Roger.

Rachel's brother Robert, ignoring her wishes to be cremated and have a simple memorial service, planned an elaborate funeral, which was held on April 17 at the National Cathedral in Washington. The honorary pall-bearers included Bob Hines, Robert Cushman Murphy, Edwin Way Teale, Senator Abraham Ribicoff, and Secretary of the Interior Stewart Udall. A large wreath of flowers sent by Prince Philip of England, an admirer of Carson's work, was a testimony to her international prominence. Two days later, Carson's close friends, including Dorothy Freeman, Marie Rodell, Jeanne Davis, and Bob Hines, gathered at All Souls Unitarian Church, where Reverend Duncan Howlett presided over the simple service that Rachel had wanted. Robert Carson later conceded to having his sister cremated but wanted the ashes buried alongside their mother in Parklawn Cemetery. Rachel, however, had wanted her ashes scattered on Southport Island. Robert compromised by keeping half of the ashes and sending half to Dorothy Freeman.

# Chapter 12

# THE LEGACY OF
# RACHEL CARSON

## THE GROWTH OF THE ENVIRONMENTAL
## MOVEMENT

Rachel Carson's legacy is reflected in the development of the environmental movement that she affected so profoundly. Patricia Hynes acknowledged Carson's impact, writing, "In every manifestation of the current environmental movement—Earth Day and the burst into being of grassroots environmentalism, ecofeminism and women's environmental activism, and the creation of the government's most vital agency, the Environmental Protection Agency—her influence is evident" (Hynes 1989: 46). *Since "Silent Spring"* author Frank Graham also recognized Carson's important role in society: "If America ever chooses to adopt a sane, coordinated conservation policy—an *environmental* policy—a great deal of the credit must go to Rachel Carson" (Graham 1970: x). Although all of her writing promoted the preservation of the environment, *Silent Spring* was the most influential of her works, motivating readers not only to appreciate nature, but also to take action to save it. In his introduction to the 1994 edition of *Silent Spring*, Vice President Al Gore asserted that the book "came as a cry in the wilderness, a deeply felt, thoroughly researched, and brilliantly written argument that changed the course of history. Without this book, the environmental movement might have been long delayed or never have developed at all" (Gore 1994: xv).

Pollution had been recognized as a problem long before the publication of *Silent Spring,* as demonstrated by the passage of the Federal Insecticide, Fungicide and Rodenticide Act in 1947, the Water Pollution Control Act of 1948, the Air Pollution Control Act of 1955, and the Clean Air Act of 1963. After Carson's death, however, the pace of the environmental movement began to accelerate, especially in regard to the pesticide problem. In July 1964, the Federal Committee on Pest Control was created to serve as "a sort of federal watchdog, creating a dialogue on pesticide problems among the departments and setting up a system of checks and balances" (Graham 1970: 188). This committee was formed to follow through with the suggestions made in the PSAC report "Use of Pesticides." Also in 1964, a federal law enacted by Congress required pesticide manufacturers to demonstrate the safety of their products before marketing them.

The Environmental Defense Fund, which filed a lawsuit in 1971 leading to the ban on DDT, was founded in 1967 by four Long Island scientists and grew to become one of the most influential environmental organizations in the country, with more than three hundred thousand members. In 1969, the National Environmental Policy Act (NEPA) was passed, establishing national guidelines for the protection of the environment. Among its other purposes was to "encourage productive and enjoyable harmony between man and his environment" and "to enrich the understanding of the ecological systems and natural resources important to the Nation" (*National Environmental Policy Act* 1969: section 4321).

## THE ENVIRONMENTAL PROTECTION AGENCY AND THE DDT BAN

On April 22, 1970, 20 million demonstrators observed the first Earth Day, a nationwide celebration that promoted environmental principles. A few months later, the Environmental Protection Agency (EPA) was formed, consolidating all environmental pursuits into one agency. The purpose of the EPA was to develop national policies that would protect the environment by regulating pesticides and other hazardous substances, but states reserved the power to enforce these regulations. Carson had recommended just such an agency during her Senate testimony in 1963. Up until that time, the Department of Agriculture had regulated pesticides, and the Food and Drug Administration set safe levels for pesticides in food. Norman Boucher suggests, "The most significant immediate effect of Carson's book—and even it took almost a decade to happen—was the establishment of the EPA and that agency's eventual banning of most chlorinated hydrocarbons—DDT, aldrin, and dieldrin, for example— from agricultural use" (Boucher 1987: 44).

In the same year that the EPA was formed, Frank Graham's *Since "Silent Spring"* was published, providing a review of Carson's life and an analysis of her most controversial work, reaction to it, and its effects on pesticide usage. In *Since "Silent Spring,"* Graham, who acknowledged that the effects of pesticides were still not understood completely, observed, "Despite the wide publicity given Rachel Carson's warnings seven years ago, man continues to expose himself to a broad range of poisons" (Graham 1970: 141). Graham shared with Carson a distrust of governmental bureaucracy, calling the federal government "myopic" and "ponderous." Although he praised men such as President Kennedy and Secretary of the Interior Udall who supported environmental causes, he believed that "the general response in Washington remains ambiguous" (185).

In 1972, the EPA cancelled the registration of DDT in the United States, which resulted in its removal from the market because all pesticides were then required to be registered with the government. An EPA press release from December 31, 1972, reported that EPA Administrator William D. Ruckelshaus "said he was convinced that the continued massive use of DDT posed unacceptable risks to the environment and potential harm to human health." Although DDT could still be used in emergency situations such as the control of disease and was, for a time, exported to other countries, the ban ultimately led to the end of DDT manufacturing in the United States. The organic phosphates aldrin and dieldrin were banned in 1974 and chlordane in 1988.

The positive effects of the DDT ban were significant. In an EPA report released in 1975, "DDT—A Review of Scientific and Economic Aspects of the Decision to Ban Its Use as a Pesticide," studies were cited that proved the human intake of DDT in the United States had decreased from 13.8 milligrams per day in 1970 to 1.88 per day in 1973. The report also indicated that DDT levels had declined drastically in a variety of fish and birds since the ban of the pesticide. Ten years after the prohibition of DDT, the *EPA Journal* reported that populations of birds such as the bald eagle, brown pelican, osprey, and peregrine falcon were increasing, and this was attributed to the ban. These birds had been on the verge of extinction because DDT had allegedly caused their eggs to have soft shells or no shells. In 1997, a quarter of a century after the EPA ban of DDT, the Environmental Defense Fund reported that the population of bald eagles had increased tenfold since 1972, from five hundred pairs in the continental United States to five thousand. One admirer of Carson contended, "Perhaps the greatest tribute to *Silent Spring* is that the tragedy Carson predicted has been averted. DDT and the other long-lived chlorinated insecticides that she warned about are banned in the United States. Species of hawks, reptiles, and fish once driven nearly to extinction are returning" (Mattill 1984: 72).

Although the formation of the Environmental Protection Agency had many positive effects, it did not ensure that risks from pesticides were obliterated. A brochure published by the Rachel Carson Council (RCC) asserts, "The use of any chemical pesticide involves risk. This is reflected in the EPA Policy that does not allow pesticide manufacturers to label their products as being safe" (Rachel Carson Council 2000). Another RCC publication clarifies that simply because a pesticide is registered with the EPA "does not mean that the chemical is 'safe' but only that when used exactly according to the label instructions, its benefits have been judged to outweigh its risks" (Rachel Carson Council 1995). Carson encouraged less reliance on government pesticide regulations, as she commented, "People say, 'We wouldn't be allowed to use things if they were dangerous.' It just isn't so. Trusting so-called authority is not enough. A sense of personal responsibility is what we desperately need" (quoted in Howard 1962: 105).

Al Gore criticized the government's efforts to regulate pesticide use as insufficient in his introduction to Silent Spring, contending, "we get short-term gain at the expense of long-term tragedy" (Gore 1994: xxi). He then affirmed that the Clinton administration would pursue stricter regulations of pesticides, limiting use, and increasing biological methods of pest control as Carson had recommended. In an effort to decrease dangers to consumers, the Food Quality Protection Act became law in 1996, requiring the EPA to review all pesticide regulations to insure a "reasonable certainty" that a pesticide will not cause harm.

Trends regarding the use of pesticides show that although Carson awakened concern about pesticide use, her warnings have gone unheeded by much of society. Banned pesticides have been replaced by other hazardous substances. Although the U.S. Department of Agriculture, for example, reduced its spraying of DDT from 4.9 million acres in 1957 to only one hundred thousand acres in 1967 and eliminated its use of this pesticide altogether by 1969, DDT was replaced, in part, by organic phosphates such as parathion, malathion, and diazinon, substances that are toxic to the nervous system and may cause behavior disorders. By 1999, there were more than twenty thousand pesticide products available, most of them toxic to a certain degree, many more than when Carson wrote Silent Spring.

According to EPA data, 617 million pounds of "conventional pesticide active ingredient" were used nationwide in 1964. This figure steadily increased until 1979 when it exceeded 1.1 billion pounds. Although the amount of pesticides used then steadily decreased, with 912 million pounds applied throughout the United States in 1999, the percentage of these pesticides used in agriculture increased from 59 percent in 1964 to

77 percent in 1999. In other words, the amount of pesticides applied to food sources nearly doubled between the publication of *Silent Spring* and the end of the century. In light of this situation, Carol Glotfelty wrote, "Considering the fact that both in the United States and overseas chemical pesticide use has increased since 1962, it would seem that the lasting significance of *Silent Spring* is to be found in attitudes more than in actions" (Glotfelty 1996: 165).

## LINGERING CONTROVERSIES

The effect of pesticides on human health was the most controversial issue in *Silent Spring* and, despite years of research since its publication, remains unresolved. Although Carson provided statistics in her book that reveal an increase in malignant cancers and an increase in the use of pesticides, her evidence was considered inconclusive because it did not demonstrate a causal relationship between the two. In 1975, the EPA acknowledged that, based on animal tests, DDT might still be a potential cause of cancer in humans. A 1989 study conducted by the National Cancer Institute and the Olin Corporation, a manufacturer of DDT, however, concluded that DDT could not be definitively proven to be a cause of human cancers.

A *Congressional Quarterly* report published in 1999 declared, "the government cannot say for certain whether pesticides, when used according to legal requirements, have no adverse effect on human health" (*CQ Researcher* 1999: 673). Earlier in this report, it was observed that "an increasing number of studies are linking pesticides on food and in the environment to a number of health disorders, ranging from breast and prostate cancer to aggressiveness and reduced motor skill ability" (667). In a study that was published in the July 2001 issue of the medical journal *Lancet,* scientists at the National Institute of Environmental Health Sciences, the National Institute of Child Health and Human Development, the University of North Carolina at Chapel Hill, and the Centers for Disease Control and Prevention found a direct correlation between DDT and premature births in the 1960s. They determined that the level of DDE, a substance produced when DDT begins to break down in the body, was elevated in the stored blood samples of mothers who had given birth to premature infants between 1959 and 1966.

Some critics of Carson have blamed her for the deaths of millions of people who have died from malaria, claiming that half a billion deaths from this disease were prevented because of the use of DDT and that hundreds of millions of people have contracted malaria since the pesticide was banned. In

December 2000, although more than a hundred countries signed a treaty with the purpose of eliminating twelve persistent organic pollutants (POPs) including DDT, twenty-five nations were allowed to continue using DDT to control mosquitoes carrying malaria. Although there are effective biological methods that have been suggested to control infected mosquitoes, numerous authorities argue that the use of DDT in developing countries to prevent this disease is justified and that safer and more effective pesticides are too expensive. As Carson predicted, however, mosquitoes are becoming increasingly resistant to the pesticides intended to eradicate them, which will eventually render these chemicals useless.

Hundreds of other species of insects have become resistant to pesticides, making it necessary to apply larger amounts of pesticides more frequently. In one study, it was determined that 31 percent of crops were lost to insects and other pests in 1945 and 37 percent in 1990, yet over the same period of time the use of pesticides increased 33 percent. In other words, a greater amount of pesticides was necessary to harvest a smaller amount of crops. Because of this situation, many scientists recommend Integrated Pest Management (IPM), a technique that reflects Carson's recommendations. IPM has been described as a "common-sense approach that uses good planning, pest monitoring, and appropriate control methods, including the judicious use of pesticides when necessary, to get the best long-term results with the least disruption of the environment" (U.S. Fish & Wildlife Service 2000).

## CARSON'S LITERARY LEGACY

Paul Brooks asserted that Carson's "meticulous research, the courage in the face of adversity and opposition, the poet's gift for langauge...made her books such an effective union of science and literature," and that "she used words to reveal the poetry—which is to say the essential truth and meaning—at the core of any scientific fact" (Brooks 1972: ix, 7). Carson believed that science writing should be included in the broad category of literature and not considered something separate. In accepting the National Book Award, she asserted, "The aim of science is to discover and illuminate truth. And that, I take it, is the aim of literature, whether biography or history or fiction. It seems to me, then, that there can be no separate literature of science" (quoted in Lear 1998: 91). Indeed, her books are studied from both a scientific and literary perspective and are considered great works of literature.

Carson was a perfectionist when it came to her writing and would agonize over each sentence, revising paragraph by paragraph rather than writing a rough draft and returning later to revise it. She was described as "a slow, self-critical writer who worked best in the hours when the rest of the world was not awake to interrupt her" (Sterling 1972: 115). She admitted to "enjoying the stimulating pursuit of research far more than the drudgery of turning out manuscript" (quoted in Brooks 1980: 280). To Rachel, writing was a very solitary pursuit, which she considered "a lonely occupation at best." She observed that "during the actual work of creation the writer cuts himself off from all others and confronts his subject alone" (quoted in Lear 1997: 286). Because of this, she believed that "only the person who knows and is not afraid of loneliness should aspire to be a writer" (quoted in Brooks 1972: 2).

Carson, who is often compared to nineteenth-century writer Harriet Beecher Stowe, can also be likened to another prominent writer of that time, Henry David Thoreau. Both Thoreau, who is best known for his account of a year spent living alone at Walden Pond, and Carson were considered reclusive, enjoyed the simple life, and had a deep appreciation of nature. A volume of Thoreau's *Journal* could usually be found at Rachel's bedside. Not only did they both write books about nature, they also wrote works that provoked controversy and sought to change society: Carson's *Silent Spring,* which criticized the misuse of pesticides, and Thoreau's "Civil Disobedience," which objected to taxes used to support war. There is, however, a distinct difference between the works of Thoreau and Carson. Placing himself at the center of his books and humanity at the center of the universe, Thoreau's writing is mostly about his own experiences. Carson, in contrast, is more of an observer, reflecting her perspective that humanity exists on the periphery of the natural world. Both writers, however, are considered to have elevated nature writing to the level of literature, and both have been linked to "a tradition of American nature writers who saw it as the writer's responsibility to interpret nature for others in order to preserve and protect it" (McCay 1993: 85). Other writers that belonged to this group include Ralph Waldo Emerson, John Muir, and Henry Beston.

## COMMEMORATING RACHEL CARSON

The memory of Rachel Carson was preserved in a variety of ways. A short time after she died, a group of Rachel's friends, including Shirley Briggs and some other colleagues, founded the Rachel Carson Trust for the Living Environment. The immediate purpose of this organization was to respond to the huge amount of mail that Carson had never had the chance to answer. Renaming itself the Rachel Carson Council, it has be-

come an international clearinghouse of pesticide information that pro-
duces a variety of publications and studies and continues the work that
Rachel Carson started to protect the environment. The mission of the organization is "to inform and advise people and institutions about the effects of pesticides that threaten the health, welfare and survival of living
organisms and biological systems." The Rachel Carson Council "promotes
alternative, environmentally benign pest management strategies to encourage healthier lifestyles" and "fosters a sense of wonder and respect toward nature." According to Shirley Briggs, executive director of the
Council for more than twenty years, the "main objective" set by the
Rachel Carson Council was "to show the ecological truths so effectively
that contending factions may in time be persuaded to join in a wider view,
and come together in common cause" (Briggs 1970: 11).

Even though Rachel Carson was often critical of the federal bureaucracy, she was ultimately recognized by the government as one of the nation's most distinguished citizens. The Presidential Medal of Freedom, the
highest award that can be given to a civilian, was awarded to Carson
posthumously in 1980 by President Jimmy Carter, who said of her, "Always concerned, always eloquent, she created a tide of environmental
consciousness that has not ebbed." The following year, a Rachel Carson
stamp was issued by the U.S. Postal Service.

Concerning her personal belief in immortality, Rachel wrote to
Dorothy a year before her death, "It is good to know that I shall live on
even in the minds of many who do not know me, and largely through association with things that are beautiful and lovely" (quoted in Freeman
1995: 446). Her memory is indeed preserved in such an appropriate way.
On the fifth anniversary of Carson's death, an article by journalist and author Ann Cottrell Free appeared in *This Week,* a supplement included in
many Sunday newspapers, suggesting that a national wildlife refuge
should be named in honor of Carson. Thousands of readers wrote to Secretary of the Interior Walter Hickel supporting this idea, and in 1969 the
Coastal Maine National Wildlife Refuge, established in 1966, was renamed the Rachel Carson National Wildlife Refuge and was dedicated
the following year. This refuge preserves thousands of acres of salt marshes
and estuaries for migrating birds and other wildlife in ten areas spread over
more than fifty miles of coastline in southern Maine.

There is another beautiful location in Maine where the memory of
Rachel Carson is preserved. On the rocky shore below Newagen Inn on
Southport Island, a plaque, succinctly describing Carson as "Writer, Ecol-
ogist, Champion of the Natural World," commemorates the place where
Dorothy Freeman scattered her friend's ashes. This was one of their fa-

vorite spots, and they frequently visited it. After the death of Rachel's beloved cat Moppet, she met Dorothy there. As they sat on a bench overlooking the ocean, they watched the migrating monarch butterflies. Later that day, Rachel wrote to Dorothy:

> But most of all I shall remember the Monarchs, that unhurried westward drift of one small winged form after another, each drawn by some invisible force. We talked a little about their migration, their life history. Did they return? We thought not; for most, at least, this was the closing journey of their lives.
>
> But it occurred to me this afternoon, remembering, that it had been a happy spectacle, that we had felt no sadness when we spoke of the fact that there would be no return. And rightly—for when any living thing has come to the end of its life cycle we accept that end as natural. (quoted in Freeman 1995: 467–68)

Rachel Carson found a deeper meaning in the migration of the monarch butterflies, reflecting her characteristic reverence for all of nature's wonders. Appreciating that all living things are interconnected, she believed that throughout nature, life is never really lost because the death of any creature ultimately leads in some way to life for others. Although Carson died at a relatively young age and would have accomplished much more if she had lived longer, she fulfilled a great purpose as her legacy continues to improve life for all the inhabitants of earth.

# BIBLIOGRAPHY

## PRIMARY SOURCES: SELECTION OF PUBLISHED WORKS BY RACHEL CARSON

### Books

1941. *Under the Sea-Wind: A Naturalist's Picture of Ocean Life*. New York: Simon & Schuster.

1951. *The Sea Around Us*. New York: Oxford University Press.

1955. *The Edge of the Sea*. Boston: Houghton Mifflin.

1961. *The Sea Around Us*. Rev. ed. New York: Oxford University Press.

1962. *Silent Spring*. Boston: Houghton Mifflin.

1965. *The Sense of Wonder*. New York and Evanston: Harper & Row.

### Articles and Chapters

1937. "Undersea." *Atlantic Monthly*, September, 322–25.

1939. "How about Citizenship Papers for the Starling?" *Nature Magazine*, June/July, 317–19.

1944. "Ocean Wonderland." *Transatlantic*, March, 35–40.

1944. "The Bat Knew It First." *Collier's*, 18 November, 24 ff.

1945. "Sky Dwellers." *Coronet*, November.

1948. "The Great Red Tide Mystery." *Field and Stream*, February, 15–18.

1953. "Mr. Day's Dismissal." *Washington Post*, 22 April, A26.

1955. "Biological Sciences." In *Good Reading*. New York: New American Library.

1956. "Help Your Child to Wonder." *Woman's Home Companion*, July, 25–27, 46–48.

1958. "Our Ever-Changing Shore." *Holiday*, July, 70–71, 117–20.

1959. "Vanishing Americans." *Washington Post*, 10 April, 26.

1960. "To Understand Biology." In *Humane Biology Projects*. Washington, D.C.: Animal Welfare Institute, 1960.

1963. "Rachel Carson Answers Her Critics." *Audubon Magazine*, September, 262–65.

1964. Foreword to *Animal Machines*, by Ruth Harrison. London: Vincent Stuart.


### Government Documents

U.S. Department of the Interior. Fish and Wildlife Service. 1943. *Food from the Sea: Fish and Shellfish of New England*. Conservation Bulletin no. 33. Washington, D.C.: Government Printing Office.

———. 1943. *Food from the Sea: Fishes of the Middle West*. Conservation Bulletin no. 34. Washington, D.C.: Government Printing Office.

———. 1944. *Food from the Sea: Fish and Shellfish of the South Atlantic and Gulf Coasts*. Conservation Bulletin no. 37. Washington, D.C.: Government Printing Office.

———. 1945. *Food from the Sea: Fish and Shellfish of the Middle Atlantic Coast*. Conservation Bulletin no. 38. Washington, D.C.: Government Printing Office.

———. 1947. *Chincoteague: A National Wildlife Refuge*. Conservation in Action Series, no. 1. Washington, D.C.: Government Printing Office.

———. 1947. *Parker River: A National Wildlife Refuge*. Conservation in Action Series, no. 2. Washington, D.C.: Government Printing Office.

———. 1947. *Mattamuskeet: A National Wildlife Refuge*. Conservation in Action Series, no. 4. Washington, D.C.: Government Printing Office.

———. 1948. *Guarding Our Wildlife Resources: A National Wildlife Refuge*. Conservation in Action Series, no. 5. Washington, D.C.: Government Printing Office.

———. 1950. *Bear River: A National Wildlife Refuge*. Conservation in Action Series, no. 8. With Vanez T. Wilson. Washington, D.C.: Government Printing Office.

# SECONDARY SOURCES

## Books

Anticaglia, Elizabeth. 1975. *Twelve American Women*. Chicago: Nelson-Hall.

Bonta, Marcia Myers. 1991. *Women in the Field: America's Pioneering Women Naturalists*. College Station: Texas A&M University Press.

Brooks, Paul. 1972. *House of Life: Rachel Carson at Work*. Boston: Houghton Mifflin.

———. 1980a. *Speaking for Nature: How Literary Naturalists from Henry David Thoreau to Rachel Carson Have Shaped America*. San Francisco: Sierra Club.

Downs, Robert B. 1970. *Books That Changed America*. New York: Macmillan.

Freeman, Martha, ed. 1995. *Always, Rachel: The Letters of Rachel Carson and Dorothy Freeman, 1952–1964*. Boston: Beacon Press.

Gartner, Carol B. 1983. *Rachel Carson*. New York: Frederick Ungar.

Glotfelty, Cheryll. 2000. "Cold War, *Silent Spring:* The Trope of War in Modern Environmentalism." In *And No Birds Sing: Rhetorical Analyses of Rachel Carson's "Silent Spring,"* ed. Craig Waddell, 157–73. Carbondale: Southern Illinois University Press.

Gore, Al. 1994. Introduction to *Silent Spring,* by Rachel Carson. Boston: Houghton Mifflin.

Graham, Frank. 1970. *Since "Silent Spring."* Boston: Houghton Mifflin.

Hynes, H. Patricia. 1989. *The Recurring "Silent Spring."* New York: Pergamon Press.

Lear, Linda. 1997. *Rachel Carson: Witness for Nature*. New York: Holt.

———, ed. 1998. *Lost Woods: The Discovered Writing of Rachel Carson*. Boston: Beacon.

———. 2000. "Afterword: Searching for Rachel Carson." In *And No Birds Sing: Rhetorical Analyses of Rachel Carson's "Silent Spring,"* ed. Craig Waddell, 205–18. Carbondale: Southern Illinois University Press, 2000.

Lutts, Ralph H. 2000. "Chemical Fallout: *Silent Spring,* Radioactive Fallout, and the Environmental Movement." In *And No Birds Sing: Rhetorical Analyses of Rachel Carson's "Silent Spring,"* ed. Craig Waddell, 17–41. Carbondale: Southern Illinois University Press.

McCay, Mary A. 1993. *Rachel Carson*. New York: Twayne.

Schweitzer, Albert. 1965. *The Teaching of Reverence for Life*. Translated by Richard and Clara Winston. New York: Holt, Rinehart, and Winston.

Sterling, Philip. 1970. *Sea and Earth: The Life of Rachel Carson*. New York: Crowell.

Whitten, Jamie L. 1966. *That We May Live*. Princeton, N.J.: Van Nostrand.

Zwinger, Ann H. 1989. Introduction to *The Sea Around Us,* by Rachel Carson. New York: Oxford University Press.

## Articles and Essays

Boucher, Norman. 1987. "The Legacy of *Silent Spring.*" *Boston Globe Magazine,* 15 March, 17, 37–47.

Briggs, Shirley. 1970. "Remembering Rachel Carson." *American Forests*, July, 9–11.

Brooks, Paul. 1980b. "Rachel Carson." In *Notable American Women: The Modern Period*, ed. Barbara Sicherman and Carol Hurd Green, 138–41. Cambridge and London: Belknap Press.

Browne, Cornelius. 2002. "Rachel Carson." In *American Writers: A Collection of Literary Biographies, Supplement IX*, ed. Jay Parini, 19–36. New York: Charles Scribner's Sons.

"Carson, Rachel Louise 1907–1964." 1992. In *Contemporary Authors, Vol. 35 (New Revision Series)*, 82–84. Detroit: Gale Research.

"Carson, Rachel (Louise)." 1951. In *Current Biography*, 100–102. New York: H. W. Wilson.

Crawford, Tom. 1995. "Rachel Carson." In *Notable Twentieth Century Scientists, Vol. 1*, ed. Emily J. McMurray, 322–25. New York: Gale Research.

Diamond, Edwin. 1963. "The Myth of the 'Pesticide Menace.' " *Saturday Evening Post*, 28 September, 16, 18.

Durgin, Cyrus, 1951. "Overnight Miss Carson Has Become Famous." *Boston Globe*, 20 July, 1, 4.

Free, Ann Cottrell. 1992. *Since "Silent Spring": Our Debt to Albert Schweitzer and Rachel Carson*. Washington, D.C.: Flying Fox Press.

Gartner, Carol B. 1979. "Rachel Carson." In *American Women Writers: A Critical Reference Guide from Colonial Times to the Present, Vol. 1*, ed. Linda Mainiero, 301–6. New York: Frederick Ungar.

Glotfelty, Cheryll. 1996. "Rachel Carson." In *American Nature Writers, Vol. 1*, ed. John Elder. 151–71. New York: Charles Scribner's Sons.

Hines, Bob. 1991. "The Woman Who Started a Revolution: Remembering Rachel Carson." *Yankee Magazine*, June, 62–67.

Howard, Jane. 1962. "The Gentle Storm Center: A Calm Appraisal of *Silent Spring*." *Life*, 12 October, 105 ff.

Interview. 1951. *Washington Post*, 4 July, 3B.

Kimler, William. 1990. "Rachel Carson." In *Dictionary of Scientific Biography*, ed. Frederic Holmes, 142–43. New York: Charles Scribner's Sons.

Lear, Linda. 1999. "Rachel Carson." In *American National Biography, Vol. 4*, ed. John A. Garraty and Mark C. Carnes, 474–76. New York: Oxford University Press.

Lewis, Jack. 1992. "Rachel Carson." *EPA Journal* (May/June): 60–62.

Mattill, John. 1984. "Looking Anew at *Silent Spring*." *Technology Review*, November/December, 72–73.

Norman, Geoffrey. 1983. "The Flight of Rachel Carson." *Esquire*, December, 472.

Pauley, Philip J. 1988. "Woods Hole and the Structure of American Biology, 1882–1925." In *The American Development of Biology*, ed. Ronald Rainger,

Keith Rodney Benson, and Jane Meienschein, 121–50. Philadelphia: University of Pennsylvania Press.

Rachel Carson Council. 1995. *A Parent Alert: Children at Risk from Household Pesticides*. Silver Spring, MD: Rachel Carson Council.

———. 2000, March. *A Nature-Lovers' Alert: Birds at Risk from Garden Pesticides*. Silver Spring, MD: Rachel Carson Council.

"Rachel Carson Dies of Cancer." 1964. *New York Times*, 15 April, 1.

"Regulating Pesticides." 1999. *CQ Researcher*, 6 August 6, 665–88.

Stinnett, Caskie. 1992. "The Legacy of Rachel Carson." *Down East*, June, 38–43.

U.S. Department of the Interior. Fish and Wildlife Service. 2000. "Homeowner's Guide to Protecting Frogs: Lawn and Garden Care." Washington, D.C.: Government Printing Office.

Walker, Martin J. 1999. "The Unquiet Voice of *Silent Spring:* The Legacy of Rachel Carson." *Ecologist*, August/September, 322–25.

Wareham, Wendy. 1986. "Rachel Carson's Early Years." *Carnegie*, November/ December, 20–34.

Watson, Bruce. 2002. "Sounding the Alarm." *Smithsonian*, September, 115–17.

Williams, Terry Tempest. 1992. "The Spirit of Rachel Carson." *Audubon Magazine*, July/August, 104–7.

## Book Reviews

Note: The following is a chronological list of book reviews cited in the text. For a comprehensive list of reviews, see "Carson, Rachel" in *Book Review Digest* published by H. W. Wilson for the following years: 1941 (*Under the Sea-Wind*); 1951 (*The Sea Around Us*); 1952 (reissue of *Under the Sea-Wind*); 1955, 1956 (*The Edge of the Sea*); 1962, 1963 (*Silent Spring*); 1965 (*The Sense of Wonder*).

Compton, A. H., et. al. 1941. *Scientific Book Club Review*, October, 1.

*New York Times*. 1941. 5 November, 27.

Sutton, G. M. 1941. *Books*, 14 December, 5.

Beebe, William. 1941. *Saturday Review of Literature*, 27 December, 5.

*Kirkus*. 1951. 1 June, 285.

Leonard, Jonathan Norton. 1951. *New York Times Book Review*, 1 July, 1.

Jackson, J. H. 1951. *San Francisco Chronicle*, 3 July, 14.

Ellis, Harry. 1951. *Christian Science Monitor*, 5 July, 7.

La Monte, Francesca. 1951. *New York Herald Tribune Book Review*, 5 July, 3.

*New York Times*. 1952. 27 April, IV: 8.

Beston, Henry. 1952. *Freeman*, 3 November, 100.

Poore, Charles. 1955. *New York Times*, 26 October, 29.

Ellis, Harry. 1955. *Christian Science Monitor*, 10 November, B: 8.

Berrill, N. J. 1955. *Saturday Review*, 3 December, 30.

Hawkes, Jacquetta. 1956. *New Republic*, 23 January, 17–18.

Milne, Lorus, and Margery Milne. 1962. *New York Times Book Review*, 23 September, 1.

Cowen, Robert. 1962. *Christian Science Monitor*, 27 September, 11.

*Time.* 1962. 28 September, 45–48.

Eiseley, Loren. 1962. *Saturday Review*, 29 September, 18.

Darby, W. J. 1962. *Chemical and Engineering News*, 1 October, 60.

Bates, Marston. 1962. *Nation*, 6 October, 202.

*Christian Century.* 1962. 19 December, 1564.

*Economist.* 1963. 23 February, 711.

## Videos

"The Silent Spring of Rachel Carson." 1963. *CBS Reports.* Produced by Jay McMullen. New York: CBS News, 3 April.

"Rachel Carson's *Silent Spring.*" 1993. *American Experience* (PBS). Written and produced by Neil Goodwin. Cambridge: Peace River Films.

## Internet Sources

Lear, Linda. 2002. "Rachel Carson and the Awakening of Environmental Consciousness." *Nature Transformed: The Environment in American History: Wilderness and American Identity.* National Humanities Center. June. www.nhc.rtp.nc.us/tserve/nattrans/ntwilderness/essays/carson.htm.

———. 2003. "RachelCarson.org." www.rachelcarson.org.

"The Power of One." 2003. U.S. Environmental Protection Agency. March. www.epa.gov/epahome/people2_0608.htm.

Rachel Carson Council, Inc. 2003. members.aol.com/rccouncil/ourpage.

Rachel Carson Homestead. 2003. www.rachelcarsonhomestead.org.

Rachel Carson Institute. 2003. www.chatham.edu/rci.

Rachel Carson National Wildlife Refuge. 2003. rachelcarson.fws.gov.

"The Story of *Silent Spring.*" 1997. National Resources Defense Council. April. www.nrdc.org/health/pesticides/hcarson.asp.

# INDEX

**About the Author**

ARLENE QUARATIELLO is a freelance writer and the author of *The College Student's Research Companion*. She has worked as an academic reference librarian at Merrimack College and Emerson College.